The
BIRDS *of* PREY
of AUSTRALIA

A field guide to Australian raptors

Stephen Debus

W0009191

Melbourne

OXFORD UNIVERSITY PRESS

Oxford   Auckland   New York

in association with

Birds Australia

OXFORD UNIVERSITY PRESS AUSTRALIA

Oxford New York
Athens Auckland Bangkok Bogota Bombay
Buenos Aires Calcutta Cape Town Dar es Salaam
Delhi Florence Hong Kong Istanbul Karachi
Kuala Lumpur Madras Madrid Melbourne
Mexico City Nairobi Paris Port Moresby
Singapore Taipei Tokyo Toronto Warsaw
and associated companies in
Berlin Ibadan

OXFORD is a trade mark of Oxford University Press

National Library of Australia
Cataloguing-in-publication data:

Debus, Stephen J. S.
The birds of prey of Australia: a field guide

Bibliography.
ISBN 0 19 550624 3.

1. Birds of prey – Australia – Identification. I. Title.

598.90994

Edited by Cathryn Game
Line illustrations by Mike Bamford, Nicolas Day,
Kate Gorringe-Smith, and Frank Knight
Colour illustrations by J. N. Davies
Text design and typesetting by Derrick I. Stone Design
Cover designed by Steve Randles
Printed by Kyodo Printing Co (S'pore) Pte Ltd
Published by Oxford University Press
253 Normanby Road, South Melbourne, Australia

Contents

Preface

Birds of prey, by virtue of their powers of flight and vision, their imposing manner and their impressive predatory capabilities, have captured the imagination of humans. Although in recent times birds of prey have not always been thought of favourably, they have long been valued as agents of the hunt in the ancient art of falconry, or as totemic figures in tribal cultures. Today, the first close encounter with a bird of prey can leave a lasting impression. To a few, like myself, it means a lifelong fascination with these magnificent birds. For me, the catalyst was the awesome sight of Wedge-tailed Eagles in flight at close range, juxtaposed with their sorry corpses strung up on paddock fences, in the Riverina of New South Wales in the 1960s. Since then I have seen all the Australian species in the field, and been privileged to know several species intimately by observing their breeding cycle and studying their biology.

Notwithstanding the animosity of certain sections of the community, mainly those concerned with pigeon-racing and livestock, birds of prey still generate much interest. They have a high profile in television documentaries and popular literature, and are increasingly in the public eye as individual birds fall foul of the dangers of the modern industrial world. Wildlife rescue services are inundated with injured, orphaned or poisoned birds. For those coming into contact with birds of prey, questions arise repeatedly: What kind is it? What does it eat? How does it live? The answers are crucial if wildlife managers and rehabilitators are to do their jobs properly.

Three decades ago, for a boy in a small country town, little relevant information was available. In the way of books there was Cayley's *What Bird is That?* and Leach's *Australian Bird Book* (both poor as field guides), a British book that said a little about European birds of prey, and a fictional book on the Peregrine Falcon in North America. The appearance of Slater's *Field Guide to Australian Birds* in 1970 was a milestone.

Today there are many good field guides, photographic books, a major reference handbook, and a treatise on the biology and ecology of Australian birds of prey. Nevertheless, good ornithologists can have difficulty identifying some raptors, even in the hand, and many find them baffling. Yet they are easy to identify if one knows what to look for, even at great distances.

My purpose in this book is to provide an inexpensive guide to the Australian raptors that will enable laypeople and bird enthusiasts alike to identify raptors, whether in the field or in the hand, and to learn a little about the birds' biology. It is intended to supplement rather than compete with the other books by, I hope, taking a fresh approach to the problem of identification.

This book draws heavily on volume 2 of the *Handbook of Australian, New Zealand and Antarctic Birds* (*HANZAB*), prepared by the Royal Australasian Ornithologists Union (RAOU) and published by Oxford University Press in 1993. Much of the information is condensed from that volume, and the colour plates of raptors in flight are reproduced here. I also gratefully acknowledge Lynx Edicions for permission to use some of the information on Australian raptors in the *Handbook of the Books of the World*, vol. 2. Original research by David Eades, Danny Rogers, and David James, section editors of *HANZAB*, is incorporated in the species descriptions.

I gratefully acknowledge the early encouragement and support of my parents, Graham and Beatrice Debus, and the later encouragement of many colleagues, including Dr Peter Jarman (my graduate supervisor at the University of New England), Dr David Baker-Gabb (RAOU Director 1992-97), and latterly my boss at UNE, Dr Hugh Ford. I thank my referees, Tom Aumann and Dr Penny Olsen, who commented helpfully on a draft of the manuscript, and Peter Higgins (co-editor of *HANZAB*) who cast an RAOU eye over it. The responsibility for any errors is mine.

Chapter 1

Birds of prey

This book offers only a brief overview of the Australian raptors. Other books that provide more detail are listed in the bibliography, along with major references and papers published since *HANZAB*.

This introduction makes a few generalisations about raptors, and a little more detail will be found in the species accounts. Each genus or group of related genera is dealt with in a separate chapter. For each genus of more than one species, a general introduction to that genus appears at the start of the chapter. An account of each species in that genus then follows, with sections on Description, Distribution, Food and Hunting, Behaviour, Breeding, and Threats and Conservation. Movements are mentioned briefly, for some species, where appropriate. The English names used here are those recommended by Birds Australia; only the most commonly used alternatives are given. I have refrained from listing folk names, in the belief that the official names should become standard.

WHAT IS A RAPTOR?

In the strict sense, a raptor is a member of the order Falconiformes—diurnal birds of prey. Included are the hawks, eagles, Old World vultures, kites, harriers, Osprey and falcons, but nowadays not the New World vultures and condors, which are related to storks and only convergently resemble the Old World vultures. The definition of *raptor* is sometimes extended to include the unrelated owls, order Strigiformes, on account of their similar structure and way of life. The owls are not considered here as they deserve a similar book of their own.

Raptors are readily characterised by their hooked bills, intense eyes, and powerful feet, each having three toes forward and one back, with sharp, curved claws. They are predominantly coloured in browns, greys, whites, and blacks, but often have brightly coloured bare parts such as yellow feet and skin around the eyes, yellow or red eyes, and yellow cere (pronounced 'seer')—the soft skin on top of the bill through which the nostrils open. Owls differ in having a flatter face (often with an obvious facial disc), more forward-directed eyes, a dull cere virtually hidden by the facial feathers, and the ability to perch with the feet arranged as two toes forward and two back.

THE KINDS OF RAPTORS

The twenty-four species of Australian raptor fall into two families: the Accipitridae—hawks, eagles, Old World vultures and allies, and the Falconidae—the falcons. The remaining raptor family, Sagittariidae, containing the Secretarybird, is found only in Africa. Australian species are native, added to which are a few vagrants from the New Guinea region. No foreign raptor species have been introduced to Australia.

The Accipitridae is divided into two subfamilies. The subfamily Pandioninae contains a single species, the piscivorous Osprey, which has special adaptations for catching fish by diving into water. The subfamily Accipitrinae contains all the other hawks, eagles, and allies, and can be conveniently divided into several informal groups of related genera. Only those found in Australia are considered here. These groups are:

1 the white-tailed kites (*Elanus*);
2 the endemic Australasian hawks, which lack close relatives in other regions but seem related to each other (the Square-tailed Kite, Black-breasted Buzzard, Red Goshawk and relatives in New Guinea);
3 the crested hawks or bazas (*Aviceda*) and honey-buzzards (*Pernis*, which do not reach Australia);
4 the large kites and sea-eagles (*Milvus, Haliastur* and *Haliaeetus*);
5 the harriers (*Circus*);
6 the goshawks and sparrowhawks (*Accipiter*);

7 the buzzards and allies, including true or 'booted' eagles with feathered legs (*Aquila* and *Hieraaetus*).

Two groups are not represented in Australia: the Old World vultures and snake-eagles. A comparison of skeletons by Richard Holdaway challenges the traditional view of the various groups, but scarcely affects the Australian members other than to split the honey-buzzards and bazas.

The Falconidae is represented in Australia only by the cosmopolitan genus *Falco*, the true falcons. The family is most diverse in South America, with genera of pygmy falcons and falconets also occurring in the Old World tropics. The major external differences distinguishing the falcons from the hawk family are the 'toothed' upper mandible (bill) with a corresponding notch near the tip of the lower mandible; black head markings varying from a full 'helmet' to a wispy malar stripe; and brown eyes surrounded by a prominent ring of yellow or pale skin. The Australian falcons, and most others in the genus *Falco*, also have long pointed wings.

IDENTIFICATION

Detailed feather-by-feather descriptions and good illustrations or photographs of Australian raptors give a very good idea of what the various species look like up close, obligingly immobile and perched and, in some cases, in flight. In practice, however, the raptor one is trying to identify is often a nondescript brown bird disappearing into the distance. Even in the hand, a raptor might look like any one of a confusing array of similar species.

A more fruitful approach is to ignore minute plumage details and instead concentrate on shape, relative proportions, flight behaviour (the way a bird holds and flaps its wings), and characteristic behaviour or mannerisms and calls. Size is also important, particularly if a reference point against which to judge relative size is present. The characteristic underwing patterns of certain species can be obvious in good views.

Look for the shape of the wings: are they long or short, broad or narrow, pointed, 'fingered' (splayed primary feathers) or rounded? Look also at the length and shape of the tail—it could be long or short, wedge-shaped, rounded, square or notched (but beware of differences in shape caused by moult of the tail feathers). The length of the legs and projection of the head and bill in front of the wings might help. Most importantly, look for the way the wings are carried in gliding and soaring flight (the positions can differ) and the style of flapping flight. The wings could be back-swept (flexed at the carpals) or straight out, or the carpals might be carried forward, producing curved leading and trailing edges. Bear in mind that the degree of flexure, and closing up of the primaries to produce pointed tips, depends on the speed and angle of gliding descent. The wings may be held above the plane of the body in a shallow or steep V when gliding or soaring, or held level with the body, or they may be bowed with the tips below the plane of the body. Look especially for the speed and depth of flapping—is it slow or fast, deep or shallow? Be careful, however, of the effect that strong wind has on wing carriage and flapping style. Finally, at close range, note any details like conspicuous plumage markings, eye colour, bare or feathered legs, and, if perched, the bird's posture and relative lengths of legs, wing tips, and tail (that is, whether the wing tips fall short of the tail tip or project beyond it). Locality, habitat and perching site will help. In the hand, note the colour of the cere, eyes, and feet; relative lengths of wings, tail, legs, and toes; nature of the legs (naked or feathered, type of scalation); and any prominent plumage markings.

Observers should be aware that rare plumage aberrations can occur in any species. These are typically leucistic (pale or 'washed-out') variations to part or all of the normal plumage, but can include scattered white feathers or all-white plumage: leucino if the bare parts (eyes, cere, feet) are normally pigmented, albino if the bare parts are also unpigmented. Other characteristics, as described, must then be used to identify such birds.

In the species accounts, total length is measured from the tip of the bill to tip of the tail. The relative length of the tail is given, to convey an impression of bodily proportions. Wingspan is the maximal natural extension from wing tip to wing tip. Most Australian species are not strongly sexually dichromatic in plumage, although all are sexually dimorphic in size, to varying degrees, with females larger than males. The mean of recorded body weights for adult-sized birds (including fledged juveniles) is given, separately for the sexes. In addition to terms such as *gliding, soaring, hovering, poising,* and *stooping,* which have precise meanings and are defined in the glossary, some explanation of flight attitudes is necessary. The term *dihedral* (having two plane surfaces) describes the V shape formed when the two wings are raised above the plane of the body. This is qualified as strong, medium or slight to denote the degree to which the wings are raised above the horizontal plane (greater than 15°, 5–15° and 0–5° respectively). A modified dihedral is when the inner part of the wing, from body to carpal, is raised and the outer part, from carpal to primaries, is mostly flat, with the primary tips curled up in those species having emarginated primaries.

Details for ageing and sexing raptors are given in the species accounts. Additionally, juveniles of many species can be recognised in the hand by their fresh, evenly worn flight and tail feathers, often with pale tips showing as a uniform, translucent trailing edge, and sometimes with symmetrical fault bars extending uniformly across neighbouring feathers. Juveniles of many hawk species also have pale tips to the greater upperwing coverts, showing as a pale line down the centre of the spread upperwing. Fledglings, just out of the nest, have the bases of their flight and tail feathers still ensheathed in 'blood' quills and may have traces of down sticking to the ends of their body feathers or showing on their head. Raptors are adult-sized at fledging, other than short wing and tail feathers; so small size does not necessarily mean that the bird is a baby. Downy chicks can be identified by family or genus characteristics such as the nature of the bill (e.g. presence of tomial 'teeth'), colour and nature of the down, character of the feet (length of tarsi and toes, type of tarsal scalation), and any distinctive markings. Raptor nestlings pass through two downy stages: the first down rather short and silky, and the second thicker and woollier (especially in falcons).

FOOD AND HUNTING

In the broad sense, many birds are predatory: they catch and eat other animals. Even small and medium-sized songbirds take some small vertebrates, such as skinks, mice, eggs, and nestlings of birds. Raptors have simply taken it a step further by catching and eating larger vertebrates, including other birds, up to or greater than their own body size, and have rather fearsome weaponry to do the job.

Raptors vary greatly in their food and feeding habits. However, all Australian species eat other vertebrates, and for some they are a major part of the diet. For many species in southern regions diet now means introduced rats, mice, rabbits or feral birds. Some, such as the Pacific Baza and Nankeen Kestrel, are mainly insectivorous. Some others, such as the large kites and Brown Falcon, eat a wide variety of prey; they and the large eagles eat some carrion. A few are highly specialised, such as the Osprey, which eats fish, the *Elanus* kites, which eat rodents, and the Square-tailed Kite, which takes small prey from the tree canopy. One, the Brown Falcon, seems particularly adapted to catching venomous snakes, and another, the Black-breasted Buzzard, uses stones to break the eggs of large ground-nesting birds, such as the Emu, and perhaps smaller eggs too.

Raptors hunt either on the wing or from a perch. Small species of open country, such as the small kites and Kestrel, often hover. The short-winged goshawks of wooded country skulk in ambush on a concealed perch and flash out to attack. Most of the falcons are highly aerial, coursing rapidly at low levels or quartering and soaring, then attacking in a stoop. Many of the larger raptors, such as kites and eagles, spend some of their time searching from slow soaring or quartering flight. The acute vision of raptors is highly sensitive to the movement of prey, but high visual resolution also permits some to see cryptic, immobile prey. In addition, harriers have an owl-like facial ruff, which might help them to pinpoint the sounds of prey in dense cover. Prey is taken by stealth and surprise. A chase develops if the prey is alerted before contact.

Raptors catch their prey by seizing or striking it with the feet, and dismember it by holding it with the feet and tearing with the bill, often starting at the head. Members of the hawk family make a kill by clutching strongly with the feet to crush the prey and drive the claws through its vital organs. The falcon family makes the kill by holding the prey in the feet and biting through the vertebrae at the base of the skull if the heavy blow from the feet, delivered in flight, has not already done the job. Either way, death comes quickly if not instantly. Captured prey is carried off in the feet, or eaten on the spot if it is small or too heavy to carry. A large meal results in a visibly bulging crop, which can change a raptor's normally sleek silhouette. Leftovers might be cached for later retrieval and consumption.

The indigestible remains of a raptor's meal, such as insect exoskeletons, bones, fur, feathers, teeth, claws, and scales, are regurgitated as a pellet or 'casting'. Pellets, along with other remains, such as feathers or bones, discarded as the raptor eats its prey, can be collected under roosts and nests and used to analyse the diet. The diets of most common species in eastern Australia, the Wedge-tailed Eagle in south-western Australia, and the Red Goshawk in the tropics have been analysed, but we still know little about the diets of the endemic species or of most species in south-western Australia, the arid zone, and the tropics.

BEHAVIOUR

The aspects of raptor behaviour that receive most attention from researchers are hunting behaviour and success, breeding dispersion and territoriality, social behaviour, and breeding behaviour. Many of these aspects have been covered fairly well for common species in eastern Australia, but there is scope for more work on hunting success in most species and on all aspects of behaviour in the uncommon or endemic species, particularly in the south-west, arid zone and tropics. For many common and well-studied species, the next step is to study social organisation and behaviour in a population of individually marked birds and to conduct community studies. Nevertheless, serious amateurs can still contribute much by taking careful notes, relating observations to prior knowledge (use *HANZAB* 2 as a starting point), and publishing significant findings.

Each raptor genus has its own characteristic way of doing things, and familiarity with behaviour patterns is also a valuable aid to raptor identification. Look for the way a raptor moves, forages and perches, and the sorts of places in which it conducts these activities. There is really no substitute for getting to know a few common species intimately.

Aerial displays are associated with breeding. They proclaim

ownership of territory and probably allow females to judge the aerial prowess of prospective mates and hence their hunting ability. The members of the hawk family employ aerial manoeuvres, such as undulating dives, slow-flapping flight with lowering of the legs, and ritualised attack and defence. The falcons employ aerobatics and ritualised attack and defence, with special postures on perches and at nest sites. These activities culminate in courtship (or supplementary) feeding, which reveals the male's capacity as a provider and enables the female to achieve the body condition necessary for laying eggs. Here, as in *HANZAB*, displays are given descriptive names such as Undulating Display (also called 'Sky-dancing'), Flutter-flight Display and Bowing Display.

BREEDING

Most Australian raptors breed in stick nests in trees or, in some species, occasionally on artificial structures. Members of the hawk family usually build their own nests, although some will also refurbish existing nests, sometimes of another species. They line their nests with green leaves or, in the large scavenging kites commensal with humans, also with human rubbish. An exception is the Swamp Harrier, which nests on the ground or in swamps. Unlike the hawks, falcons do not build their own nest but adopt an old stick nest or even usurp one from the original owner. Some, such as the Australian Hobby, remove the lining. The Brown Falcon occasionally adds material to an existing nest. The Kestrel and Peregrine Falcon also breed in tree hollows and on ledges of cliffs or city buildings.

In general, clutch size varies with body size, ranging from one or two eggs in large eagles to five or six in the small *Elanus* kites and Kestrel. The length of each phase of the breeding cycle also varies with body size. Incubation ranges from four weeks in the smallest species to six weeks in large eagles. Similarly, nestling periods range from four weeks in the smallest species to almost three months in large eagles.

Raptors usually breed in spring, but in some species the laying season extends throughout the dry season, from the austral autumn to spring, in the moist tropics. Most species rear a single brood in any given year. However, the Black-shouldered Kite breeds between

Osprey *Pandion haliaetus* **1** Adult male; **2**, **3**, **4** Adult females; **5**, **6**, **7** Juveniles

Head of a female Osprey *(Nicholas Birks)*
Juvenile Osprey in flight *(Nicholas Birks)*

Black-shouldered Kite stretching its wings *(Nicholas Birks)*

Letter-winged Kite *(Nicholas Birks)*

Pacific Baza *Aviceda subcristata* **1** Adult female; **2** Adult male; **3** Juvenile; Black-shouldered Kite *Elanus axillaris* **4** Adult; **5** Juvenile; Letter-winged Kite *Elanus scriptus* **6** Adult male; **7** Adult female; **8** Juvenile

Black-breasted Buzzard *Hamirostra melanosternon* **1** Adult; **2** Juvenile, fresh plumage; **3** First immature; **4** Second immature; Square-tailed Kite *Lophoictinia isura* **5** Adult; **6** Juvenile, fresh plumage; **7** Juvenile, worn plumage

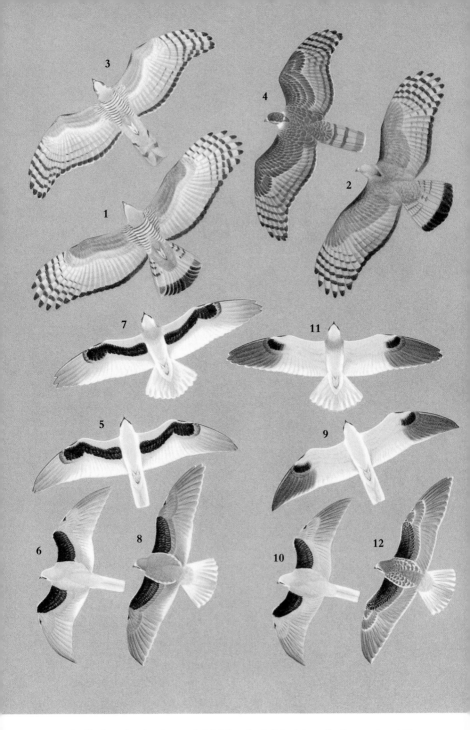

Pacific Baza *Aviceda subcristata* **1**, **2** Adult males; **3**, **4** Fresh juveniles; Letter-winged Kite *Elanus scriptus* **5**, **6** Adults; **7**, **8** Juveniles; Black-shouldered Kite *Elanus axillaris* **9**, **10** Adults; **11**, **12** Juveniles

Black-breasted Buzzard *Hamirostra melanosternon* **1**, **2** Adults; **3**, **4** Juveniles, fresh plumage;
5, **6** Juveniles, worn plumage; **7** Second immature; Square-tailed Kite *Lophoictinia isura*
8, **9** Adults; **10**, **11** Juveniles

A Black-breasted Buzzard in flight *(Nicholas Birks)*
Black Kites gathering *(Nicholas Birks)*

autumn and spring, sometimes rearing two broods in a year. The Letter-winged Kite breeds continuously during rat plagues.

Most Australian species breed in simple, dispersed pairs. Notable exceptions are the Letter-winged Kite, which nests in loose colonies, and the Black-breasted Buzzard, which very occasionally, under some circumstances, breeds in polyandrous trios (although this behaviour requires investigation). Male goshawks and sparrowhawks are occasionally bigamous, and some harriers are polygynous, although this is rare in the Swamp Harrier, which is usually monogamous. In most raptors the male's role is to provide the food from courtship and laying until the nestlings are well grown, when the female is then freed to hunt. In many species the male will cover eggs or chicks while the female is off feeding on prey provided by him. The male might even share daytime incubation. In a few, notably the Square-tailed Kite and Black-breasted Buzzard, males take a greater share in nest duties.

The usual division of labour between the sexes is intimately connected with the phenomenon of reversed sexual size dimorphism, in which female raptors are larger than their mates. It is particularly noticeable in raptors that hunt difficult prey (large or agile birds) in situations that place the hunter at risk of injury, either from the prey or from collisions with obstacles. It seems that females have evolved relatively large body size either to compete successfully for a good provider in a good territory or as a hedge against periods of food shortage while they are tied to nesting duties. This subject requires further study.

The breeding biology of most common species in eastern Australia, and of the Wedge-tailed Eagle in Western Australia and Red Goshawk in the tropics, has been well studied. A comprehensive account of the breeding cycle has been given for several other species, also in the east. Again, the gaps in knowledge are greatest for the endemic and other uncommon species, most species in the south-west, arid zone and tropics and, notably, for the common but little-studied Black-shouldered Kite. For all Australian species we have at least a general idea of the usual breeding season, site and construction of the nest, clutch size, incubation and nestling periods, and parental roles. For some, breeding densities and the above-mentioned aspects have been quantified, and for many, a detailed account of events from courtship and nest-building to fledging exists. The difficulty is to obtain detailed information on the rest of the cycle, from the post-fledging period to independence of the young. We still have too little information on breeding success, in terms of the number of young fledged per territorial pair per year in sample locations, for many species.

Today the buzz-words are *lifetime reproductive success* and *population viability analysis* and the associated concept of the 'metapopulation'. The latter concept applies particularly to forest-dependent species threatened by habitat destruction and the possible effect of habitat fragmentation on dispersal between subpopulations of a metapopulation. These subjects require sophisticated study of regional variation over time of aspects of population dynamics, such as density, fecundity, mortality, and dispersal, using individually marked birds. Nevertheless amateurs should contribute where they can, either by developing their own observational studies (and publishing the results) or by participating in scientific projects run by others.

HANDLING RAPTORS

Anyone with an interest in birds could be asked to rescue a raptor in distress. This subject is treated in greater detail by other books, to which readers are referred (see bibliography). The important points are that captured or cornered raptors can inflict severe damage with their feet to human hands, arms or faces, and some (notably falcons) will also bite. They will be calmed and restrained if enveloped in a dark cloth or similar when catching them, and then held in a dark, cool and ventilated container, such as a cardboard box. They panic and damage themselves severely if enclosed by wire mesh.

Handling of raptors for research purposes (banding and other forms of marking) is best learnt under the tutelage of an experienced person, and indeed permits to trap and mark raptors, or to hold and rehabilitate raptors, are (or should be) granted only to those with sufficient experience and accreditation. The subject of raptor rehabilitation is also a specialised field, dealt with in more detail in other books (see bibliography). It demands a high degree of skill, dedication and the appropriate facilities, in order to avoid problems such as further injuring or stressing captive raptors, releasing ostensibly healthy raptors to certain death, or imprinting young raptors on humans. Raptors are not pets nor suitable subjects for the average wildlife rescuer. The holding of raptors for falconry, in the sense of catching and training raptors for hunting other animals for sport, is illegal in Australia.

THREATS AND CONSERVATION

For each species I have summarised its national and global conservation status, using the accepted international terms for level of threat as determined by the definitive publication on the subject (*Threatened and Extinct Birds of Australia*, RAOU Report 82, revised 1993), and as adopted by state and federal legislation. In some cases, as indicated, the categories have been revised in *Birds to Watch 2, The World List of Threatened Birds* (BirdLife International, 1994), although Australian legislation has not caught up with these changes. In discussing the main threats, I have used the past tense in relation to the effect of DDT on the thickness of eggshells, because DDT was banned from broadscale agricultural use in Australia in 1987, and eggshell thickness of affected species is returning to normal. DDT-induced eggshell thinning increased the likelihood of egg breakage, embryo death, and poor reproductive success in affected species.

Chapter 2

Osprey *Pandion haliaetus*

In the genus *Pandion* (the name of two kings of Athens in Greek mythology) there is a single species, the cosmopolitan Osprey. It is a large aquatic hawk, with adaptations for catching fish by plunge-diving into water. It has dense plumage to avoid wetting; a large oil gland for waterproofing its plumage when it preens; no aftershafts on the feathers of its head and underparts; and powerful feet with rough spicules on the soles, long curved claws, and the ability to swivel its outer toe into a two-by-two arrangement like that of the owls. It also has flexible joints that enable it to sweep its wings clear of the water and so take off almost vertically. Like the sea-eagles, the Osprey has bare tarsi, but the pattern of scalation is reticulate, not scutellate.

DESCRIPTION

The specific name simply means 'sea eagle', which is somewhat confusing as the Osprey is a distinctive hawk not closely related to the true sea-eagles of the genus *Haliaeetus*.

The Osprey is 50–66 cm long (tail less than half), with a wingspan of 149–168 cm. The weight of males averages 1013 g, females 1235 g. Intermediate in size between the Brahminy Kite and White-bellied

Sea-Eagle, it is a large hawk with long, angular wings and heavy feet. At rest it has a distinctive profile: crested nape, bulging chest but seemingly concave belly, prominent carpals, and wing tips extending beyond the tip of the tail. The **adult** is brown with white head and underparts, a dark stripe through the eyes and down the sides of the neck, and a mottled brown breast band (narrow and faint in males, heavy in females). The underwings are faintly barred, with dark carpal patches. The tail is barred. The cere is grey, the eyes pale yellow to orange-yellow, and the feet pale grey. The **juvenile** is similar but generally rustier, with the crown and nape more streaked, the upperparts spotted cream (lost with wear), the breast band heavy and broad, and the eyes orange-yellow to orange. The **chick** is uniquely patterned among raptors, having grey-brown down with dorsal stripes and a dark eye stripe encircling the nape.

The Osprey is a solitary hawk of inshore coastal and estuarine waters, and occasionally inland rivers and lakes. Its flight is kite-like and buoyant with shallow, gentle rhythmic wing beats; it soars and glides on bowed wings with curved leading and trailing edges (see figure 2.1). Its silhouette shows a prominent head and bill, long narrow wings, and square tail. Its voice is distinctive: plaintive or loud ringing whistles, repeated.

The Osprey is most likely to be confused with the immature White-bellied Sea-Eagle, some individuals of which may show a mottled breast band on otherwise pale underparts, but that species has broad, rounded, upswept wings and a shorter, wedge-shaped tail. The juvenile Brahminy Kite has shorter, broader, rounded wings, a rounded tail, less projecting head and bill, and small feet, and soars on flat wings.

Figure 2.1 Soaring and gliding

DISTRIBUTION

In Australia the Osprey can be found around almost the entire coast-line, although it is rare in Victoria and absent from Tasmania. It sometimes occurs far inland, on rivers or lakes, particularly in wet years.

FOOD AND HUNTING

The Osprey eats mostly fish, occasionally taking crustaceans, reptiles, small mammals, or birds. It forages by soaring or quartering and sometimes hovering, or rarely by still-hunting from a perch. It dives headlong into water with the feet thrown forward, submerging in a plume of spray and with the wings raised. If successful, it flies off with a large fish slung torpedo-fashion beneath the body, the forward foot gripping the head and the trailing foot gripping the rear. Occasionally it catches seabirds in flight. The catch is carried to a prominent perch or to the nest, which is used as a feeding platform outside the breeding season.

BEHAVIOUR

Apart from its fishing behaviour and conspicuous perching on prominent sites, readily observed elements of the Osprey's behaviour are its aerial displays. Either alone or in the company of his soaring mate, a male Osprey performs undulating dives with shrill squealing calls. Sometimes, during such manoeuvres, he carries a fish in what is called the Fish Display. Members of a pair also perform following or pursuit flights on a weaving course. Once settled on a breeding territory, adult Ospreys are sedentary.

BREEDING

The Osprey lays from April to July in northern Australia and July to September in southern Australia. The nest is a large bowl or pile of sticks and driftwood up to 2 m across and 2 m deep, lined with vegetation and flotsam, in the exposed top of a dead or partly dead tree, on a cliff or on an artificial structure, from ground level (on rocks or islets in remote areas) to 45 m above the ground. The clutch size is usually three eggs, ranging from two to four. Incubation takes 35–37 days, and the nestling period is 54–60 days. Juveniles remain dependent on their parents for two or three months after fledging, then disperse widely (up to 400 km has been recorded).

THREATS AND CONSERVATION

The Osprey is not globally or nationally threatened. It is generally common and secure in Australia, particularly in the north. It is uncommon and local in the south, with its range contracted in the extreme south. The small New South Wales population of seventy or so pairs is threatened but now stable or increasing, with some pairs nesting on artificial platforms. Locally, in the south and east, pairs are threatened by disturbance and destruction of breeding habitat and nest sites near coastal developments. Direct persecution, through illegal egg-collecting, shooting, and the deliberate destruction of nest sites, seems largely to have abated. The thickness of the Osprey's eggshells was not significantly reduced by DDT use in Australia.

Chapter 3

Small kites, genus *Elanus*

The 'white-tailed' kites in the genus *Elanus* ('kite') are small, gull-like, grey-and-white hawks with black forewing patches and varying amounts of black on the underwings. They have long pointed wings, hover frequently and prey mainly on rodents, which they often swallow whole. They have short legs, stout feet, and a reticulate pattern of tarsal scalation. Of the two Australian species, the Black-shouldered Kite most closely resembles others in the genus, particularly the White-tailed Kite (*Elanus leucurus*) of the Americas rather than the geographically closer Black-winged Kite (*Elanus caeruleus*), which occurs in Africa and Eurasia through to New Guinea. The endemic Letter-winged Kite is unique among raptors in being largely nocturnal, and has the most distinctive plumage of the genus. As a group, these kites seem geared to a 'boom and bust' strategy, with a high reproductive output and extreme mobility enabling them to find concentrations of prey and rapidly build up local populations. In lean times, their populations disperse, and many birds perish.

Black-shouldered Kite
Elanus axillaris

DESCRIPTION

The specific name, part of a direct quotation from the original description in Latin, refers to the black markings on the wings.

The Black-shouldered Kite is 33–37 cm long (tail less than half), with a wingspan of 82–94 cm. The weight of males averages 249 g, females 293 g. It is slightly larger than the Nankeen Kestrel, with a larger head, broader wings, and shorter tail. The **adult** is a small white hawk with pale grey back and wings, darker grey primaries, and a black patch on the upperwings between the carpal flexure and body. It has a black carpal spot and dark grey primaries on the underwings. There is a black patch in front of, and a thin black line above and behind, the eyes. The cere is horn to yellow, the eyes are red, and the feet rich yellow. The **juvenile** is washed or streaked rusty brown on the head, back, and breast. Feathers of the upperparts, including the black wing patch, are fringed white; the cere is horn, the eyes brown, and the feet pale yellow. The **chick** has fawn first down and pale grey second down, which is slightly browner on the crown.

The Black-shouldered Kite is a solitary or gregarious hawk of open woodland and grassland. It is characteristic of farmland with scattered trees. At rest it sits low on short legs, with the wing tips extending beyond the tip of the tail. Its flight action is rapid and winnowing. It glides on raised wings (see figure 3.1), hovers with the legs lowered and tail depressed, and drops feet-first with the wings raised high over the back (compare with Nankeen Kestrel). After landing on a perch, it sometimes flicks the tail upwards while

Figure 3.1a Soaring and gliding

Figure 3.1b Gliding

Figure 3.2 Tail-wagging

giving weak whistling notes (see figure 3.2). Its other common call is a harsh wheezing.

The Black-shouldered Kite is similar to the Letter-winged Kite except for the underwing pattern, facial pattern, and style of flight (for differences, see the account for the Letter-winged Kite). It is more robust, shorter-tailed, and cleaner white underneath than the Nankeen Kestrel, and lacks the black band on the tail tip. The Grey Falcon lacks the black markings on the forewings and underwings, has barred wings and tail, and is an active bird-hunter. The Grey Goshawk has short broad rounded wings without black markings, has longer tail and legs, and glides on bowed wings.

DISTRIBUTION

The endemic Black-shouldered Kite occurs throughout Australia, least commonly in the arid zone and Tasmania.

FOOD AND HUNTING

The Black-shouldered Kite eats mostly small rodents. Occasionally it takes small birds, lizards, and insects. It is diurnal and crepuscular, and occasionally nocturnal on moonlit nights. It forages by quartering and hovering, or sometimes by still-hunting from a perch, then dropping on to prey on the ground.

BEHAVIOUR

The Black-shouldered Kite's most readily observed behaviour is its hovering and its conspicuous perching on utility poles and wires and on the topmost branches of live or dead trees. It also roosts communally. Individuals harass larger raptors aggressively. Members of a pair engage in aerial chases and the Flutter-flight Display, with aerial food-passes by the male to the female (see figure 3.3). Occasional Talon-grappling occurs, either between rivals or in courtship.

Figure 3.3 Aerial transfer of food

BREEDING

The laying season varies. It extends throughout most of the year but has peaks in autumn and spring. Some pairs might have two broods a year. Pairs nest solitarily or in loose colonies when prey is abundant. The nest is a platform of sticks 27–45 cm across and 13–15 cm deep, lined with green leaves and placed 4–35 m above the ground in the canopy of a live tree or, rarely, on an artificial structure. The clutch size is usually three or four eggs, ranging from two to five. Incubation takes about 31 days, and the nestling period is 33–38 days. Juveniles remain dependent on their parents for about a month, then disperse widely (up to 1000 km has been recorded). They might breed when a year old.

THREATS AND CONSERVATION

The Black-shouldered Kite is not nationally threatened. It is widespread and common over much of its range, and has increased in range and numbers in cleared and farmed areas of southern Australia in response to creation of suitable habitat and introduction of suitable prey (the House Mouse *Mus domesticus*). The thickness of its eggshells was not significantly affected by DDT use in Australia. Locally, secondary poisoning might occur when rodenticides are used during mouse plagues or pesticides are used during locust plagues.

Letter-winged Kite
Elanus scriptus

DESCRIPTION

The specific name ('written') refers to the black M or W marking across the underwings.

The Letter-winged Kite is 34–37 cm long (tail less than half), with a wingspan of 84–89 cm. Males weigh 289 g on average, females 343 g. It is similar in size, colour, and proportions to the Black-shouldered Kite, although with softer and more owl-like plumage.

Adults are white with pale grey back and wings, and a black patch on each wing between the carpal and body. Some birds have a grey wash on the crown and nape, or even a grey 'cap'; males are whiter on the crown. The underwings have a thick black line from the body to the carpal joints, and pale grey primaries (the reverse of the under-wing pattern of the Black-shouldered Kite). The Letter-winged Kite has a black patch in front of each eye joining a black ring around the eye, which, with the large eyes, enhances the owl-like facial expression. The cere is dark horn to black, the eyes red, and the feet cream. The **juvenile** is washed brown on the head, back and breast, lacking the white feather-tipping of the juvenile Black-shouldered Kite (other than a thin white line along the outermost scapulars). It has brown eyes. The **chick** has cream first down and light grey-brown second down, which is darker on the head.

The Letter-winged Kite is a gregarious kite of tree-lined water-courses and adjacent grasslands of the eastern arid zone. Its flight action is slow and harrier-like, with gliding on raised wings (see figure 3.4); also sustained wind-hanging on motionless wings raised above the body. Its hunting behaviour is similar to that of the Black-shouldered Kite, but its wing-beats are slower and deeper. Calls consist of weak whistling and harsh rasping notes (like the calls of the Black-shouldered Kite) and a harsh slow chatter.

The Letter-winged Kite is distinguished from the Black-shouldered Kite by its underwing pattern and paler, more translucent flight feathers, as well as by its flight, which is slower, more buoyant, and less direct. It is similar to the Barn Owl (*Tyto alba*) and

Figure 3.4a Soaring and gliding

Figure 3.4b Gliding

Grass Owl (*Tyto capensis*) at night, but those species have larger heads, broader rounded wings, and long dangling legs.

DISTRIBUTION

The Letter-winged Kite normally lives in the eastern interior of Australia, with its distribution centred on the Barkly Tableland and Channel Country. Following occasional wet seasons and a return to normally dry conditions, birds irrupt into atypical habitats and localities, including open areas on the coast and even offshore islands. Some extralimital breeding occurs during rat and mouse plagues, in habitats structurally resembling its normal habitat.

FOOD AND HUNTING

The Letter-winged Kite eats mostly rodents, especially the native Long-haired or Plague Rat *Rattus villosissimus* during plagues; it also takes other small mammals, reptiles, and insects. It is normally nocturnal, like its main prey. It forages by quartering and hovering, then dropping on to prey on the ground. During irruptions, starving birds out of their normal range still-hunt from low perches.

BEHAVIOUR

The Letter-winged Kite usually roosts communally by day in well-foliaged trees. Birds sometimes roost along rather than across branches. During periodic irruptions towards the coast, when birds are starving, they show aberrant hunting and roosting behaviour, such as diurnal activity and roosting on the ground. Breeding colonies are silent and inactive by day, with sentinels perching quietly above nests. If disturbed, birds soar high over the site and give alarm calls. Members of a pair engage in the diurnal Flutter-flight Display, circling high with wings held in a tight V. Food passes from male to female are often performed in flight.

BREEDING

The laying season is irregular. It is continuous (mostly autumn to spring) during times of abundant prey following rains, but there is little breeding in years between plagues of rats. Pairs breed in loose colonies but occasionally solitarily. The nest is a platform of twigs and herbage 28–51 cm across and 20–34 cm deep, lined with green leaves or cattle dung, and placed 2–11 m above the ground in the canopy of a live tree. The clutch size is usually four or five eggs, ranging from three to six. Incubation probably takes about 31 days, and the nestling period is 30–35 days. During plagues of rats, clutches and broods are large, and new clutches can be laid while fledglings are still on the previous nests.

The period of dependence after fledging is apparently short. Birds mature rapidly and are possibly able to breed within their first year if favourable conditions persist, leading to rapid growth of breeding colonies. When conditions deteriorate, young disperse widely from breeding colonies in the Cooper Creek drainage system to all Australian coasts. Most involved in such irruptions are first-year or second-year birds, which die. Repopulation of the normal range is probably by a small nucleus of surviving adults that remain.

THREATS AND CONSERVATION

The Letter-winged Kite is not nationally threatened, although it is of special concern. It is generally uncommon. Its core range and population are small and subject to habitat degradation by over-grazing, which exacerbates the effects of regular droughts. Breeding colonies are also threatened by the increase in numbers of feral cats, which occupy nests and are suspected of eating the young. Rodenticides might cause secondary poisoning during mouse plagues, when the Kites invade agricultural areas.

Chapter 4

Australian endemic hawks, genera *Lophoictinia, Hamirostra, Erythrotriorchis*

As first pointed out by Penny and Jerry Olsen, the three rather distinctive hawks in this group appear to be related to each other and to several New Guinean hawks that are also in their own endemic, and in some cases monotypic, genera. The Australian species are the Square-tailed Kite, Black-breasted Buzzard and Red Goshawk. The New Guinean species are the long-tailed buzzards (genus *Henicopernis*, which closely resembles *Lophoictinia*), Doria's Hawk (*Megatriorchis doriae*, which somewhat resembles *Erythrotriorchis*), the Chestnut-shouldered Goshawk (*Erythrotriorchis buergersi*, a dusky, short-winged, more piebald version of the Red Goshawk), and perhaps the New Guinea Harpy-Eagle (*Harpyopsis novaeguineae*, resembling a giant Doria's Hawk in some respects). Penny Olsen has suggested that this assemblage of related species might conceivably include the Philippine Eagle (*Pithecophaga jefferyi*). They are probably ancient relics of a bygone age when Australia was part of the southern supercontinent, Gondwana. These species appear to have no close relatives on other continents, but their distant relatives might be found on other Gondwanan remnants (Africa or South America). Perhaps taxonomists should look to two other isolated, monotypic genera: the Madagascar Serpent-Eagle (*Eutriorchis astur*) and another African 'old endemic', the Congo Serpent-Eagle (*Dryotriorchis spectabilis*). Those two little-known species have not been placed convincingly with the 'core' snake-eagles (*Circaetus*) and serpent-eagles (*Spilornis*).

During long isolation as Australia drifted towards Asia, the endemic hawks probably radiated to fill vacant ecological niches and, in so doing, convergently evolved to resemble the true kites, buzzards and goshawks (some of which later invaded Australia from

Asia). These special Australasian hawks are therefore as unique as the Malleefowl, lyrebirds and bowerbirds. The Australian representatives of this hawk assemblage are characterised by much rufous in the plumage, especially in juveniles, and strongly patterned underwings. As a group, the 'old endemics' seem to be primitive stock, now uncommon to rare, declining in the face of colonisation by 'modern' raptors, and, most significantly, the drastic habitat changes wrought by modern humanity.

Square-tailed Kite
Lophoictinia isura

DESCRIPTION

The generic name ('crest kite') refers to the small occipital crest, and the specific name ('equal tail') refers to the square-cut tip to the tail. The Square-tailed Kite is an endemic, monotypic genus characterised by a small, slender bill, bulbous cere, lack of a bony brow ridge, small feet, and bare tarsi with reticulate scaling.

This kite is 50–56 cm long (tail about half), with a wingspan of 131–145 cm. A juvenile male weighed 501 g; females average 650 g. The Kite is similar in size to the typical or milvine kites (*Milvus*, *Haliastur*) and harriers (*Circus*) but has longer and more splayed primaries, giving a wider wingspan. It is a slender, very long-winged soaring hawk, in structure and character like the Pacific Baza (barred primaries, flight action) and like the harriers (low sailing flight on raised wings) and Black-breasted Buzzard. It is distinctive in flight, showing a white cap (in adults), wings broadest near the tips, and a long square or notched, sharp-cornered tail, which is twisted and fanned when soaring into the wind.

The **adult** has dark brown upperparts with a white forehead and face, rufous dark-streaked nape, and pale upperwing band; some birds have a pale lower back and rump. The tail is grey-brown with a dark terminal band. The underparts are rufous with heavy dark streaks on the breast. The underwings have rufous leading edges, barred secondaries (with terminal band broadest), and pale bases to the primaries (visible on the upperwings in flight); the pattern is like that of a pale Little Eagle but more diffuse, with a buffier tone, black carpal crescents, and longer, barred (not black) primary tips. The cere is pinkish white, the eyes pale yellow, and the feet pinkish white. The **juvenile** has dark brown upperparts with rufous scalloping, broadest on the upperwing band, and a thin rufous line along the centre of the spread upperwing. The head and underparts are rufous with fine dark streaks. The tail grades from completely barred on the central feathers to almost unbarred on the outermost feathers. The cere is pinkish white, the eyes brown, and the feet cream. Plumage fades to a duller, paler version with wear. **Immatures** (second-year birds) are intermediate between adults and juveniles, with darker upperparts, redder underparts and paler (although not white) head than faded and worn juveniles. The **chick** has white down, which is long and hairlike on the head.

The Square-tailed Kite is a solitary hawk of open forest, woodland, scrub, and heath in coastal and subcoastal areas and riverine trees in the inland. It is buoyant and agile in flight, sailing and circling low over or through the treetops, jinking sideways between the crowns of trees. It glides on raised wings, with the carpals held forward, primaries back-swept, although widely splayed, and a slight raising and lowering of the dihedral angle (see figures 4.1 and 4.2); it dips with a steep dihedral to snatch prey from foliage. It seldom flaps; its flight action is variously either loose and shallow or deeper with fluid, rowing wing-beats. Its gliding flight is more stable and eagle-like than that of the typical kites, sometimes with a sideways rocking motion like that of the Black-breasted Buzzard. At rest it

Figure 4.1 Soaring

Figure 4.2 Gliding

sits low on short legs with the tarsi hidden by the thigh feathers; it shows a slightly crested nape and long primaries extending well beyond the tail tip (the crossed primary tips sometimes creating the illusion of a forked tail). It seldom lands on the ground, and then usually only briefly. It is usually silent, but utters a hoarse or plaintive yelp and, near the nest, a rapid chitter or rattle.

The Square-tailed Kite can be confused with several species. The juvenile or immature Black-breasted Buzzard is more robust, with larger bill and feet, shorter black (not barred) primaries, shorter unbanded tail, and more prominent pale panels in the wings. The Whistling Kite is sandier brown with large pale areas on the upperwings and a rounded tail, and glides on bowed wings. The pale form of the Little Eagle is chunkier, with shorter primaries and tail, crisper underwing pattern, and feathered tarsi, and glides on level wings. The harriers have unbanded primaries, rounded or wedge-shaped tail tip, long legs sometimes lowered in flight, and no upperwing band or panels in the wings, and flap their wings more often. The Red Goshawk is more robust, with shorter, more pointed wings, heavier, quicker flight, and massive legs. The Square-tailed Kite ought not to be confused with the Black Kite, which is a social scavenger with black (not barred) primaries and flat or bowed wings in gliding flight. Nevertheless, Black Kites showing a square tip to the fanned tail are often misidentified as Square-tailed Kites.

DISTRIBUTION

The Square-tailed Kite is found over most of the Australian mainland and some larger offshore islands. It avoids the most arid, treeless central regions, where it is very scarce or absent. In the far south-east it is seen mostly in spring and summer, and in the far north mostly in the dry season.

FOOD AND HUNTING

The Square-tailed Kite eats mostly eggs and nestlings of birds but also adult birds; often honeyeaters and other passerines that build nests in foliage. It also eats insects, reptiles, tree-frogs and, rarely, small mammals, but does not feed on carrion. Foraging is solitary and mostly aerial by means of low, slow quartering of the tree and shrub canopy; it also hawks flying insects. From low sailing flight, it drops suddenly to snatch prey from foliage or seize prey startled into flight. It also reaches into, tears apart or removes birds' nests from trees or shrubs, and removes nests of paper-wasps in order to extract the larvae.

BEHAVIOUR

The Square-tailed Kite's most characteristic behaviour is its thorough searching of the vegetation canopy from low sailing flight, often in long transects or wide meandering arcs. Its gracefulness and agility in the air, or among the treetops, are noteworthy. At rest or in low overhead flight it is remarkably confiding and approachable, almost lethargic. In the breeding season, solitary birds perform an aerial Undulating Display, and the members of a pair perform languid following, chasing or mock attack-and-parry manoeuvres, with dodging and rolling. The species is evidently a long-distance migrant; southerly breeding birds or their offspring move north to winter in the tropics and return the following spring.

BREEDING

The laying season is from July to December. Pairs nest solitarily. The nest is a platform of sticks 50–85 cm across, 25–60 cm deep, lined with green leaves, and placed 8–34 m above ground in the fork of a living tree within forest or woodland. The clutch size is two or three eggs, usually three. Incubation probably takes about 40 days, and the nestling period is about 59–65 days. Broods of one or two young fledge. The period of dependence after fledging lasts for one or two months.

THREATS AND CONSERVATION

The Square-tailed Kite is globally threatened, classified as Rare (recently revised to Vulnerable). This highly specialised raptor is uncommon and declining in its southern and eastern breeding range, affected by habitat clearance and illegal egg-collecting. It has a low breeding density and recruitment rate. The population is estimated at twenty to fifty pairs in Victoria and fewer than ten birds in South Australia, where few recent successful breeding records exist. It seems to be adapting to the well-vegetated outer fringes of cities in subtropical eastern Australia, where it feeds on the abundant introduced and native passerines. The thickness of its eggshells was not significantly reduced by the use of DDT. Conservation measures required include a population survey and research into its biology and ecology, especially in south-eastern Australia.

Black-breasted Buzzard
Hamirostra melanosternon

DESCRIPTION

The generic name ('hook bill') refers to the prominent upper mandible, and the specific name ('black breast') describes one of the bird's most characteristic features in adulthood. The Black-breasted Buzzard is an endemic, monotypic genus, characterised by a rather long bill, bulbous cere, slight brow-ridge, robust feet, and bare tarsi with reticulate scaling.

This bird is 51–61 cm long (tail about a third), with a wingspan of 147–156 cm. The average weight of males is 1196 g, females 1330 g. It is a large soaring hawk intermediate in size between the Little Eagle and Wedge-tailed Eagle, with upswept wings and a conspicuously short square tail. In structure and character (large wings, remarkable buoyancy) it resembles the Wedge-tailed Eagle and Square-tailed Kite. The adult is unmistakable: mainly black from below with contrasting white wing panels (like the Dollarbird *Eurystomus orientalis*) and a pale tail. Juveniles and immatures are more rufous or brown.

The **adult** is mainly black, browner on the wings, and mottled rufous on the back and shoulders. The nape, thighs, and undertail coverts are rufous. The tail is grey-brown and unmarked. The underwings have a mottled brown leading edge, unmarked grey secondaries, and black primaries with white bases that form the wing panels (also visible on the upperwings in flight). The cere is pale pink, the eyes brown, and the feet pale pink. The **juvenile** has blackish brown wings with broad rufous scalloping, and a thin rufous line along the centre of the spread upperwing; the head, back and underparts are rufous with dark streaks. The tail is grey-brown. The underwings have a rufous leading edge, faintly barred secondaries, and black primaries with creamy white bases. The cere is light blue-grey to bluish white, the eyes light brown, and the feet grey-white. **Immatures** (second- or third-year birds, the so-called pale phase) are intermediate between juveniles and adults, browner than the juveniles, with the amount of black (particularly streaks on the breast) increasing over time. The **chick** has white down, which is long and hair-like on the head.

The Black-breasted Buzzard is a solitary hawk of wooded and open habitats, particularly riverine woodlands and forests, in northern and inland Australia. Its flight is buoyant, consisting of rapid low sailing or high soaring on long, broad, raised wings. The bird raises and lowers the dihedral angle and cants from side to side (see figures 4.3 and 4.4). The wings are a rather even width and held in a modified dihedral, with the primaries somewhat swept back. Active flapping is a powerful rowing action or (in pursuit) more rapid, shallow wing-beats. Its gliding flight is stable and eagle-like. It gathers in small numbers at carrion or eggs of large ground-nesting birds. At rest it shows a prominent bill, occipital crest, short legs, and primaries extending well beyond the tip of the tail. Its most common call is distinctive: a repeated hoarse yelping.

The adult Black-breasted Buzzard ought not to be confused with other species. A distant adult Wedge-tailed Eagle is black with some white in the underwings, but has a distinctive long wedge-shaped

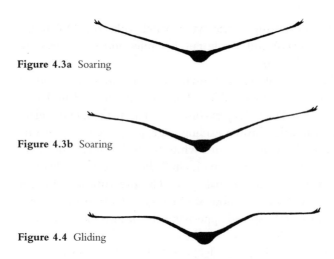

Figure 4.3a Soaring

Figure 4.3b Soaring

Figure 4.4 Gliding

tail. The dark form of the Little Eagle is duller with a pale upper-wing band, indistinct pale panels in the wings, longer barred tail, and feathered tarsi, and soars on flat wings. The juvenile or imma-ture Buzzard can be confused with several species. The Square-tailed Kite is more slender with longer barred primaries, pale upperwing band, dark carpal crescents on the underwings, and a longer tail; the Red Goshawk has shorter, less upswept wings, barred primaries, longer barred tail, massive legs, and quicker flight; the juvenile Brahminy Kite glides on flat (not upswept) wings and has a rounded tail.

DISTRIBUTION
The Black-breasted Buzzard is found over most of the Australian mainland except the temperate high-rainfall regions of the south and south-east.

FOOD AND HUNTING
The Black-breasted Buzzard eats mammals, birds, reptiles, carrion and, occasionally, large insects, particularly young rabbits, bird nestlings (including other raptors) and eggs of birds, and large lizards. It forages by flying fast gliding transects at moderate height and by low, slow quartering or high soaring. It swoops to snatch prey from

trees or seize it on the ground. It sometimes walks on the ground, searching for prey. The Buzzard will also break eggs of large ground-nesting birds with blows of its bill, or use the bill to hurl stones at the eggs from a standing position on the ground (see figure 4.5).

Figure 4.5 Breaking eggs

Behaviour

The Black-breasted Buzzard's most characteristic behaviour is its effortless soaring and sailing flight across inland skies, in long transects or wide sweeping arcs rather than the tight spiralling of most other raptors. In the breeding season, members of a pair soar over their nesting territory for hours and occasionally perform an Undulating Display or mock attacks with rolling and parrying. The Buzzard is notable for its stone-throwing behaviour, being one of only two raptors, and very few birds, that regularly use tools to assist in their feeding.

Breeding

The laying season is from June to November, usually August to October. Pairs nest solitarily. This species is usually monogamous, but there is one record of a polyandrous trio persisting for four years. The nest is a large, flat platform of sticks 70–120 cm across, 40–55 cm deep, lined with green leaves, and placed 6–22 m above ground in the fork of a living or dead tree. The clutch size is usually two eggs, less often one, rarely three. Incubation probably takes about 40 days, and the nestling period is about 60 days (see figure 4.6). Brood size at fledging is usually only one, rarely two. The period of dependence after fledging lasts between two and five months. Sexual maturity is attained at two years or older; birds occasionally breed in immature (second- or third-year) plumage.

Figure 4.6 Shading chicks

THREATS AND CONSERVATION

The Black-breasted Buzzard is not nationally threatened, but its conservation status is of concern. It is generally uncommon, and populations have declined in the south-east of its range as a result of clearance and degradation of habitat, the decline or extinction of native animals on which it fed, and the poisoning of carcasses. However, it might have benefited from the introduction of rabbits and the abundance of stock carrion. It has a low breeding density and recruitment rate, and is sensitive to human activity near the nest. It is sometimes killed when scavenging on roads, and is subject to illegal egg-collecting.

Red Goshawk
Erythrotriorchis radiatus

DESCRIPTION

The generic name is a composite usually taken to mean 'red bird of prey', the second part translating literally as 'three testicles'. Apparently this was an ancient allusion to the high level of sexual activity of raptors, which copulate frequently when nesting. The specific name ('radiated') refers to the boldly striped or barred plumage. The genus is endemic to Australasia, with a second species in New Guinea, the Chestnut-shouldered Goshawk (*E. buergersi*). It is characterised by a robust bill, slight brow ridge, and exceptionally heavy feet with bare tarsi having the scutellate pattern of scalation front and back (the scutes become fused in adulthood).

The Red Goshawk is 46–61 cm long (tail about half), with a wingspan of 111–136 cm. Two males averaged 635 g; one female weighed 1110 g and another at least 1370 g. The female is similar in size to the Little Eagle but with more tapered wings and a longer tail. The male is similar in size to a female Brown Falcon but with slightly shorter wings. The Red Goshawk is a large, active, rufous-brown hawk, boldly mottled and streaked, with a square-tipped tail, massive legs and feet, and boldly barred underwings. The long wings and loose, streaky plumage are unlike those of the true goshawks (*Accipiter* species).

The **adult male** has dark brown upperparts, streaked paler on the head and broadly scalloped rufous on the back and upperwings. The face is grey and finely streaked black. The flight feathers are slate-grey with dark bars. The tail feathers are slate-grey, barred darker; some birds lack bars on the central rectrices except for a dark subterminal bar. The underparts are rufous, brightest and unmarked on the thighs, and streaked black on the breast and flanks. The throat is pale with dark streaks. The underwings have a rufous leading edge and pale remiges with bold dark bars and crisp black tips. The cere and orbital skin are pale blue-grey to pale grey, the eyes brown to yellow, and the legs and feet yellow. The **adult female** is more heavily built than the male, with a deeper bill and thicker tarsi. She is duller and less rufous, with heavily barred central tail feathers. The underparts are off-white with dark streaks and rufous flanks; some birds have a pale rufous chest. The thighs are rufous and unmarked. The eyes are golden to pale yellow. The **juvenile** has redder upperparts (broader and more extensive rufous scalloping) than the adults and a barred tail. The head and underparts are rich rufous with fine dark streaks. The underwings are obscurely barred, with diffuse long dark tips to the primaries. The cere and orbital skin are pale blue, the eyes dark brown, the legs and feet pale grey or cream to yellow. The **chick** has white down, which is long and hair-like on the head.

The Red Goshawk is a solitary, secretive hawk of coastal and sub-coastal forest and woodland in the tropics and subtropics. Direct flight is sometimes leisurely, rather heavy and crow-like, with sustained flapping (especially by the female). Rapid flight, as in pursuit, is powerful and energetic with deep, fluid wing-beats, like a fast-flying Brown Falcon. The Red Goshawk sails and soars with parallel and almost straight leading and trailing edges to the wings, slightly bulging secondaries, and 'fingered' tips; it sometimes soars with the legs lowered. Descending glides are on flat or slightly bowed wings, with flexed carpals and pointed tips (see figures 4.7 and 4.8). It sometimes stoops with closed wings. The flat head, deep bill (female), broad deep chest, and large pale feet are often obvious. At rest it shows a deep bill (female), slight crest, powerful shoulders, long legs with large feet (long toes), and tail tip just beyond the wing tips. The voice is distinctive: raucous shrieks and cackling like the Brown Falcon and Peregrine Falcon.

Figure 4.7 Soaring

Figure 4.8 Gliding

The Red Goshawk is most likely to be confused with the rufous morph of the Brown Falcon, which is slimmer, with a double cheek-mark, rounded shoulders, more tapered wing tips (reaching the tail tip at rest), rounded tail, and smaller feet (short toes). Many other reddish-brown raptors are commonly misidentified as the Red Goshawk, in particular the Square-tailed Kite and female Swamp Harrier, which have similar plumage patterns from below; and the juvenile Black-breasted Buzzard and Spotted Harrier, juvenile or dark form of the Little Eagle, and the Whistling Kite. The Red Goshawk differs from those species in proportions and flight behaviour: the Square-tailed Kite has a small bill, much longer wings, short legs and small feet, and slow sailing flight; the harriers are slimmer with rounded or wedge-shaped tail tips, long slender legs, and slow sailing flight low to the ground; the Black-breasted Buzzard has longer wings, much shorter tail, and short legs; the Little Eagle has shorter wings and tail and feathered tarsi; the Whistling Kite has small, pale legs and feet (for further differences, see the accounts for those species). The Red Goshawk lacks the conspicuous pale markings on the upperparts (upperwing band, pale rump) of other raptors; instead, it has strikingly variegated (scalloped) upperparts and rufous thighs. The juvenile Brown Goshawk has rufous edges to the feathers of the upperparts, but has white, coarsely streaked and barred underparts, shorter wings, and a rounded tail extending well beyond the wing tips at rest.

DISTRIBUTION
The Red Goshawk is confined to a narrow coastal and subcoastal strip from the Kimberley region of Western Australia, across the Top End of the Northern Territory and Gulf of Carpentaria, to eastern Queensland and extreme north-eastern coastal New South Wales. Occasional wandering birds reach central Australia.

FOOD AND HUNTING
The Red Goshawk eats mostly birds, particularly parrots and pigeons but sometimes herons, waterfowl, kookaburras, and megapodes. It rarely takes mammals (flying-foxes, young hares), reptiles (snakes, lizards), and large insects. Early and late in the day the Goshawk forages by short-stay perch hunting from concealed positions in trees and, in the middle of the day, by long gliding or flying transects and low quartering through or just above the tree canopy, or by high soaring flight. It often seizes prey in flight after a stealthy glide or direct flying attack that becomes a vigorous chase; it also stoops on prey from a height.

BEHAVIOUR
The Red Goshawk is difficult to observe and seldom seen, other than around an active nest. It usually skulks in trees, but can be observed opportunistically as it flies past or soars high. Solitary birds or pairs soar and perform aerial manoeuvres, including an Undulating Display and mock attack-and-parry with evasive dives or rolls. In the breeding season, the male also performs low-level acrobatic flights about the perched female. Noisy food passes from male to female take place on perches near the nest, sometimes with ritual mounting by the male.

BREEDING

The laying season is from May to October in the north, August to October in the east. Pairs nest solitarily. The Red Goshawk is usually monogamous, although one male was possibly bigamous with females in neighbouring territories. The nest is a platform of sticks 60–120 cm across, 30–50 cm deep, lined with green leaves, and placed 15–29 m above ground in an exposed fork of the tallest emergent living or partly dead tree. The clutch size is one or two eggs, usually two. Incubation takes about 40 days, and the nestling period is 51–53 days for males (probably slightly longer for females). Juveniles remain dependent on their parents for at least 2–3 months and might remain in their natal territory for 4–5 months. Red Goshawks probably reach sexual maturity at two years or older, as there are no records of breeding in juvenile plumage.

THREATS AND CONSERVATION

The Red Goshawk is globally threatened and classified as Vulnerable (recently revised to Endangered). It is scarce, with specialised requirements, and locally restricted within its continental range. In eastern Australia, populations have declined and breeding range contracted through loss of habitat. It has a low breeding density and recruitment rate. It is threatened by deforestation, illegal egg-collecting, and locally by disturbance from bird-watchers and photographers at nests. It was probably affected locally by DDT use, but there are insufficient data to detect changes in eggshell thickness. The population is estimated at about 350 breeding pairs, most of which are in northern Australia.

Conservation measures taken include a biological study in northern Australia and a proposed management plan. Additional conservation measures required include monitoring of known territories in northern Australia; further surveys in eastern Australia to locate breeding territories, which should be secured and monitored; and research into the species' biology and ecology in eastern Australia.

Chapter 5

Bazas, genus *Aviceda*

The bazas or cuckoo-hawks in this genus ('bird killer', a misnomer) are distributed from Africa to Asia and Australasia. The name *baza* is also a misnomer, derived from a Hindi word for goshawk. Bazas are medium-sized, long-winged hawks lacking the bony brow-ridges, and hence the fierce expression, of most other raptors. They have a crest on the nape, small weak feet, boldly patterned or barred plumage, densely feathered lores, and specialised bills with a double tomial 'tooth' on each side of the upper mandible. The tarsus is short, with scutellate scaling in front and reticulate scaling behind. The large, bulging eyes, giving a wide field of view, enable them to detect well-camouflaged prey among foliage, and the dense facial feathering offers some protection against dangerous insects. The toothed bill might help to dismember insects and is used to bite through the skulls of reptiles and amphibians. The Australian species, which most closely resembles the Cuckoo Hawk *A. cuculoides* of Africa, has a structure and character (prominent eyes, barred plumage, short legs) reminiscent of a cuckoo. Its flight is buoyant and graceful on broad, rounded wings, making the bird look larger than it really is.

Red Goshawk *Erythrotriorchis radiatus* **1**, **2** Adult females; **3**, **4** Adult males; **5** Adult male, worn; **6** Juvenile female; **7**, **8** Juvenile males, fresh; **9** Juvenile male, worn

Whistling Kite *Haliastur sphenurus* **1** Adult; **2** Juvenile; Brahminy Kite *Haliastur indus* **3** Adult;
4 Juvenile; **5** First immature; Black Kite *Milvus migrans* **6** Adult; **7** Juvenile; Little Eagle
Hieraaetus morphnoides **8** Adult, dark morph; **9** Adult, dark morph, darkest birds; **10** Adult,
light morph; **11** Juvenile, dark morph; **12** Juvenile, light morph

Little Eagle *Hieraaetus morphnoides* **1** Adult female, dark morph; **2** Adult male, light morph;
3 Adult female, light morph; **4** Juvenile, dark morph; **5** Juvenile, light morph; Brahminy
Kite *Haliastur indus* **6** Adult; **7** Juvenile; **8** First immature

Whistling Kite *Haliastur sphemurus* **1** Adult, light; **2** Adult, intermediate; **3** Adult, dark;
4 Juvenile; Black Kite *Milvus migrans* **5** Adult, dark; **6** Adult, pale; **7** Juvenile

Whistling Kite *Haliastur sphenurus* **1** Adult; **2** Juvenile, fresh; **3** Juvenile, worn; Little Eagle
Hieraaetus morphnoides **4** Adult; **5** Juvenile; Black Kite *Milvus migrans* **6** Adult; **7** Juvenile;
Brahminy Kite *Haliastur indus* **8** Adult; **9** Juvenile; **10** Immature

Adult female White-bellied Sea-Eagle in flight *(David Whelan)*

A Spotted Harrier collecting nest material *(Nicholas Birks)*

White-bellied Sea-Eagle *Haliaeetus leucogaster* **1**, **2** Adults; **3**, **4** Juveniles, fresh plumage;
5, **6** First immatures; **7** Second immature; **8**, **9** Young adults with varying amounts of
retained immature plumage

White-bellied Sea-Eagle *Haliaeetus leucogaster* **1** Adult; **2** Juvenile; **3** First immature;
4 Second immature

Pacific Baza (Crested Hawk)
Aviceda subcristata

DESCRIPTION

The specific name ('somewhat crested') refers to the small erectile crest on the nape, which enhances a cockatoo-like appearance as the bird jumps or clambers among tree branches.

The Pacific Baza is 35–46 cm long (tail about half), with a wingspan of 80–105 cm. The weight of males averages 307 g, females 347 g. It is similar in size to the male Brown Goshawk, but has larger wings. The **adult** is slate-grey with a brown wash on the wings and lower back, a pale belly with bold dark bands, and pale rufous thighs and undertail coverts. The underwings have pale rufous leading edges, boldly barred primaries, and a broad dark trailing edge on the secondaries. The undertail has fine basal barring and a broad dark terminal band. The cere is blue-grey, the eyes bright golden-yellow to (rarely) reddish orange, and the feet pale blue-grey to cream. The **juvenile** has dark brown upperparts with pale rufous edges to the feathers, giving a finely scalloped appearance. It has pale eyebrows, a partial rufous collar, and a rufous wash over the breast. Fine rufous bars might be visible on the breast and between the bold bars on the belly. The cere is pale yellow, the eyes pale grey to pale yellow, and the feet cream to pale yellow. The **chick** has short white down, its distinguishing character being the double-toothed bill; developing dark feathers of the crest are evident while it is still downy.

The Pacific Baza is a solitary or gregarious hawk of forest, woodland, and urban trees in the tropics and subtropics. In flight the head and neck appear somewhat pigeon-like; the carpals are held forward so that the leading and trailing edges of the wings are curved; the wings are broadest at the tips and narrowest near the body, and the primaries are held widely splayed. The tail is long and square at the tip. Its flight action is loose and shallow or deeper with fluid, rowing wing-beats; the head is often held above the line of the back. It soars and glides on flat or slightly bowed wings (see figure 5.1), but raises the wings in a V at the start of a display flight. Non-breeding birds are often in groups. Undulating display flight is distinctive: the

Figure 5.1a Soaring and gliding

Figure 5.1b Soaring and gliding

Figure 5.2 Undulating display

bird climbs with deep, laboured beats then drops with the wings held in a steep V, like a pigeon (see figure 5.2). Its reedy, two-note whistle, rising and falling, is also distinctive.

The plumage of the Pacific Baza in some ways resembles that of the Brown Goshawk and Collared Sparrowhawk, but is more boldly barred. At rest it sits low on shorter legs, with the longer wing tips almost reaching the tail tip and both extending well below its perch; in flight the wings are longer and wing-beats slower. Its flight silhouette is similar to that of the typical, or milvine, kites, and confusion is also possible with the Grey Goshawk and Little Eagle. However, its wing shape and boldly barred, widely splayed primaries are diagnostic. It can be confused with birds other than raptors, such as the rainforest Topknot Pigeon (*Lopholaimus antarcticus*), which has narrower, more pointed wings, faster wing-beats, and a longer tail.

DISTRIBUTION

In Australia the Pacific Baza occurs in a coastal band from the Kimberley in northern Western Australia to the central coast of New South Wales, and rarely south of Sydney. It is largely restricted to within about 400 km of the coast, in the east occurring inland to the western slopes of the Great Dividing Range and adjoining plains. It also occurs from Wallacea through New Guinea to the Solomon Islands.

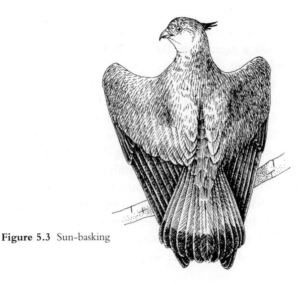

Figure 5.3 Sun-basking

FOOD AND HUNTING

The Pacific Baza is omnivorous, eating mainly foliage insects and tree-frogs but also snakes, lizards, small nestlings and birds, and small fruits such as figs. It forages by low quartering over the canopy, or still-hunting within the canopy, inspecting the foliage. Sometimes it clambers acrobatically among branches, hanging with flailing wings. It also glides or drops to snatch prey from the foliage. It is said to find tree-frogs by imitating their calls and so inducing them to call back, but this suggestion requires confirmation.

BEHAVIOUR

The Pacific Baza is an unobtrusive and docile hawk, usually obviously only during its vocal aerial displays. Patient watching and listening in likely habitat will usually be rewarded by the sight of a bird moving within or between crowns of trees, the patter of falling insect fragments, or the calls of a soaring bird. In the breeding season, pairs perform Undulating Displays and mutual soaring with much calling; they allopreen; and the male feeds the female at the nest. Variations of aerobatic displays include somersaulting at the apex of undulations, lateral rolls, side-slipping, and pursuit flights through the canopy. The Baza sometimes perches in the sun, basking with its wings open (see figure 5.3).

BREEDING

The laying season is from September to January in Australia. Pairs breed solitarily. The nest is a small platform of twigs 25–40 cm across and 12–15 cm deep, lined with green leaves, and placed within the tree canopy 6–35 m above the ground. The clutch size is usually two or three eggs, occasionally four. Incubation takes 29 days, and the nestling period is 32–35 days. Storms destroy some nests and eggs or young. The period of post-fledging dependence is not known, but some family groups persist almost to the next breeding season. In other cases, juveniles seem to form post-breeding flocks quickly, which, in south-eastern parts of the Australian range, migrate either to the lowlands or northwards for the winter before returning in spring.

THREATS AND CONSERVATION

The Pacific Baza is not globally or nationally threatened. It is common in the tropics, although uncommon at the extremities of its Australian range. It probably benefits from fragmentation of continuous forest, as it prefers edges, although it is absent from extensively cleared areas. It is possibly affected by pollutants such as lead in heavily urbanised areas. The thickness of its eggshells was not significantly affected by DDT use in Australia.

Chapter 6
Large kites and sea-eagles

The typical or milvine kites (in Australia the genera *Milvus* and *Haliastur*) and the sea- and fish-eagles (in Australia the genus *Haliaeetus*) form a closely related and discrete group, united by the possession of a fused second joint of the inner toe. The kites are characterised by easy, lazy soaring flight, with tilting and twisting of the tail; small feet; and scavenging and piratical ways of life. The sea-eagles are larger, more predatory and more strictly aquatic, with larger feet and shorter tails. Though well within, and in some cases exceeding, the size range of the true ('booted') eagles, the sea-eagles are readily distinguished by their bare tarsi.

The Australian species in the group are the near-cosmopolitan Black Kite, the near-endemic Whistling Kite, the Brahminy Kite and the White-bellied Sea-Eagle, both of which extend to South-East Asia. Sanford's Sea-Eagle (*Haliaeetus sanfordi*), a close relative of the White-bellied Sea-Eagle, is endemic to the Solomon Islands. An extinct species formerly inhabiting some South Pacific islands might have been most closely related to the huge Steller's Sea-Eagle (*Haliaeetus pelagicus*) of the Siberian coast.

Species in the group are often vocal, with loud and distinctive calls and, in the sea-eagles, characteristic duetting by mated pairs. This group, more than any other, is traditionally credited with spectacular whirling or cartwheeling courtship displays in which two birds fall through the air with locked claws. However, a recent survey has shown that most talon-grappling and cartwheeling by raptors (including sea-eagles) is fighting.

Black Kite (Fork-tailed Kite)
Milvus migrans

DESCRIPTION

The generic name simply means 'kite' (in the raptorial sense), and the specific name refers to spectacular mass migrations of milvine kites seen in Europe. The genus is characterised by long forked tails and short tarsi with scutellate scalation in front and reticulate scaling behind.

The Black Kite is 47–55 cm long (tail about half), with a wingspan of 120–139 cm. The average weight of males is 554 g, females 626 g. It is a medium-sized, scruffy brown soaring hawk that often gathers in flocks at carrion, refuse, and fires. Usually seen in slow circling flight, it shows a prominent pale band on the upperwings, and the shallowly forked or square-tipped tail (when widely fanned) is often twisted from side to side.

The **adult** is dark brown with a pale forehead and throat, a pale diagonal band on the upperwings, and black wing tips. The underparts are tinged chestnut with fine dark streaks; the underwings are barred, with black outer primaries and pale inner primaries; and the undertail is barred. The cere is yellow, the eyes brown or sometimes hazel, and the feet yellow. The **juvenile** is paler, with sandy streaks on the head, upperparts and underparts, and a bolder underwing pattern (pale bases to the primaries); it has a pale line along the centre of each spread upperwing. The cere is yellow to greenish yellow, the eyes brown, and the feet pale yellow. The **chick** has white first down, which is long and spiky on the head with a dark patch around each eye, and pale fawn second down.

The Black Kite is a gregarious hawk, loosely commensal with human settlement, and is often seen in large flocks around towns, rubbish dumps, stockyards, piggeries, abattoirs, and roads in northern and inland Australia. Solitary birds occasionally reach southern coastal areas. Its flight action is loose, with rowing wing-beats and some body movement, somewhat like the flight of a tern; it soars and glides on flat or slightly arched wings, often in a hunched, head-down posture with the carpals held forward, although the attitude of wings and body is constantly changing (see figure 6.1). The wings might be momentarily upswept when gaining height. It is highly manoeuvrable, side-slipping to snatch food from ground or water. Low pursuit flight, such as during territorial conflicts, is faster, with shorter, faster wing-beats and pointed, back-swept wings reminiscent of a Brown Falcon. At rest the Black Kite sits low on short legs, with the wing tips just reaching the tail tip. Calls are various peevish squeals, mews and whinnies; one call is like a subdued, tremulous version of the Whistling Kite's call but does not ascend.

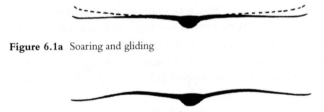

Figure 6.1a Soaring and gliding

Figure 6.1b Soaring and gliding

The Black Kite is slimmer, darker, and more agile than the Whistling Kite, with narrower, more pointed wings, a narrower band on the upperwings, and a forked (not rounded) tail. The dark morph of the Little Eagle is more robust with larger bill, head and feet, shorter primaries, shorter unforked tail, and feathered tarsi. The juvenile Brahminy Kite has broader wings and a shorter rounded tail, and lacks the upperwing band. The Black Kite is sometimes misidentified as the Square-tailed Kite, which is quite distinct (for differences, see the account for that species).

DISTRIBUTION

The Black Kite occurs mostly in northern and inland Australia, sporadically and sparsely in the south; occasionally it irrupts in areas outside its normal range. It also occurs from most of the Old World to Indonesia and New Guinea.

FOOD AND HUNTING

The Black Kite eats a variety of small animals and carrion: mammals, birds, reptiles, amphibians, fish, invertebrates, road-killed vertebrates, large carcasses and offal. Human food scraps are scavenged from camping grounds, picnic areas, sports events, and schoolyards. The Kite forages by high soaring and quartering; it drops on prey, hawks flying insects, snatches prey from water surfaces, and sometimes baits fish. It patrols roads and fire fronts, robs other raptors, and follows herds of stock or farm machinery for flushed prey. Small food items are eaten on the wing.

BEHAVIOUR

The Black Kite's most characteristic behaviour is its effortless circling in inland or tropical skies, in flocks sometimes numbering hundreds or even thousands. It can ascend beyond the range of human vision, or suddenly appear overhead, having descended from invisible heights. Also characteristic is its association with human refuse and domestic animals. It has learnt to deal with road-killed Cane Toads by turning them over and opening the belly, thus avoiding the dorsal poison glands. Individuals sunbathe and dustbathe communally on the ground and roost communally in trees. Courtship displays consist of spiralling and leisurely pursuit or tandem soaring, followed by the pair descending to the nest and calling from it, where they sometimes allopreen.

BREEDING

Laying occurs in all months in northern Australia and from July to November (rarely autumn) in the south. Pairs are solitary or loosely colonial. The nest is a platform of sticks 50–75 cm across, 60 cm deep, lined with dry vegetation, wool, fur, dung or rubbish, and placed within the tree canopy 2–30 m above the ground. The clutch size ranges from one to four eggs, usually two in poor seasons and three or four in good seasons. Incubation takes about 31 days, and the nestling period is 37–44 days. Success has been measured as 82% hatching success and 64% fledging success, and as 1.5 young fledged per clutch and 0.9 fledged per territorial pair. The period of dependence after fledging possibly lasts about two months. Other aspects of the breeding cycle are poorly known in Australia, although well studied elsewhere.

THREATS AND CONSERVATION

The Black Kite is not globally or nationally threatened. It is common to abundant in Australia, and most numerous in areas of human influence, although very scarce in the far south and scarce away from human activity in the arid zone. It has probably increased in range and numbers since European occupation. The thickness of its eggshells was not significantly reduced by DDT use in Australia.

Whistling Kite
Haliastur sphenurus

DESCRIPTION

The generic name ('sea goshawk') alludes to the aquatic, often coastal, habits of these kites, and the specific name ('wedge tail', a

slight misnomer) refers to the rounded tail tip. The genus is characterised by short legs with bare tarsi, which have scutellate scalation
in front and reticulate scaling behind, and sharp spicules on the pads
under the toes (more developed in the Brahminy Kite).

The Whistling Kite is 51–59 cm long (tail about half), with a
wingspan of 123–146 cm. Males weigh 632 g on average, females
907 g. It is similar in size to the Little Eagle but with longer wings
and tail and a slimmer head and body. It is a medium-sized, scruffy,
sandy-brown hawk, which is often attracted to carrion. In slow circling flight on slightly bowed wings with long splayed primaries, it
shows a large pale area on the upperwings and a long, rounded, pale
tail, contrasting with the dark remiges.

The **adult** is a mottled or streaky sandy brown with a pale head,
darker wings and black primaries. It has a distinctive underwing pattern with a pale leading edge, dark secondaries, pale inner primaries,
and dark outer primaries. Bars on the underwings and tail are faint
or absent. The cere is pale grey, the eyes brown, and the feet cream.
The **juvenile** is darker (rustier) than the adult, its underparts are
more streaked, and the upperparts are spotted buff or white; a pale
line along the centre of each spread upperwing is formed by pale
tips to the greater coverts. The cere is dark grey, and the eyes and
feet are as in the adult. The **chick** has cream first down, browner
on the head where it is long and spiky, and pale fawn second down,
which is browner on the back.

The Whistling Kite is a solitary or gregarious kite of most terrestrial habitats except dense forest, and is often seen around water,
including estuaries, coastlines, and inland drainages. Its flight action
is a rather deep, jerky rowing action with some body movement. It
glides on bowed wings with the carpals held forward, producing
curved leading and trailing edges to the wings (see figure 6.2). At
rest it sits low on short legs, with the long tail projecting well beyond

Figure 6.2 Soaring and gliding

the wing tips. Its voice is distinctive: a loud whistle sliding down the scale, preceded or followed by a rapid up-scale chatter.

The Whistling Kite can be confused with several species. The juvenile Brahminy Kite is more compact, with shorter primaries, wholly dark upperwings, and shorter tail, and it glides on flat wings. The Little Eagle is more robust, with a larger head, heavier feathered legs, narrower pale upperwing band, straighter wings with shorter primaries, and a shorter, square tail. It glides on flat wings. The Black Kite is darker, with a narrower pale upperwing band, narrower, more pointed wings, and forked tail. The juvenile Square-tailed Kite is more rufous, with longer barred primaries, upswept wings, and square-tipped tail.

DISTRIBUTION
The Whistling Kite occurs throughout Australia, although rarely in Tasmania. It also occurs in New Guinea and New Caledonia.

FOOD AND HUNTING
The Whistling Kite eats a variety of small animals and carrion: mammals, birds, reptiles, fish, crustaceans, insects, road-killed vertebrates, large carcasses, and offal. It forages by quartering or high soaring, and by waiting in trees beside water, especially at drying pools in creeks; it drops on prey, hawks flying insects, or snatches prey from water surfaces. It patrols roads (especially early in the morning) and fire fronts, and robs other raptors. It is rather more robust and predatory than the Black Kite.

BEHAVIOUR
The Whistling Kite's most characteristic behaviour is its lazy soaring, often accompanied by loud whistling calls. Individuals or pairs also call vigorously from prominent perches or the nest, in an upright posture with the head raised and bill tilted skywards. Aerial displays consist of mutual soaring, aerobatics, and mock attack-and-parry, with calling, and sometimes with one bird carrying food or nesting material. Allopreening and presentation of food or nesting material occur at the nest. Feeding Kites guard against theft by mantling (see figure 6.3).

Figure 6.3 Mantling over food

BREEDING

The laying season is long and variable: most of the dry season (austral autumn to spring) in the tropics, and usually late winter and spring, although also in autumn, in the south. Pairs nest solitarily. The nest is a bowl of sticks 60–150 cm across, 30–100 cm deep, lined with green leaves, and placed 3–62 m above the ground in the fork of a live or dead tree. The clutch size is usually two eggs but ranges from one to four. Incubation takes about 35 days, and the nestling period is 44–54 days. Success has been variously measured as 60% hatching success and 1.0 young raised per nest; 1.5 young per clutch started and 1.3 young fledged per pair per year; and 96% fledging success in nests with young. The period of dependence after fledging lasts up to two months, after which birds disperse, some more than 2000 km.

THREATS AND CONSERVATION

The Whistling Kite is not globally or nationally threatened. It is common to abundant on coasts and in the tropics, where it benefits from human activity. It is scarce away from water in inland Australia, and in the arid zone is largely restricted to sites with water. It is declining locally in southern Australia through drainage of wet-lands and reduction in its food supply. The thickness of its eggshells was reduced by the use of DDT in agricultural areas of southern Australia. Some birds, because of their scavenging habits, are poisoned by agricultural chemicals or baited carcasses.

Brahminy Kite
Haliastur indus

DESCRIPTION

The scientific name denotes a 'sea goshawk' from India, where it was first described. The generic characters are as for the Whistling Kite, but the Brahminy Kite has more prominent spicules on the undersides of the feet.

The Brahminy Kite is 45–51 cm long (tail less than half), with a wingspan of 109–124 cm. The average weight of males is 536 g, females 588 g. It is a medium-sized soaring hawk, slightly smaller than the Whistling Kite, with shorter wings and tail. The adult is a striking chestnut and white: unmistakable. The juvenile is mostly brown, paler on the head and underparts, without a pale upperwing band. In adults and juveniles a translucent pale tip to the tail, broadest on the central feathers, can create the illusion of a square tail.

The **adult** is mostly rich chestnut with a white head, breast and tip to the tail, and black outer primaries. The cere is cream to pale blue, the eyes reddish brown, and the feet cream. A newly fledged **juvenile** has dark brown upperparts with cream spots; mottled brown head and underparts, with a dark horizontal line through the eye; pale belly and undertail; and an underwing pattern reminiscent of a pale Little Eagle or immature Black-breasted Buzzard, although more diffuse. Juvenile plumage fades with wear to brown on the upperparts and pale on the head and underparts. The cere is grey, the eyes brown, and the feet cream. **Immature** birds are a patchy mixture of faded juvenile plumage and a dull adult-like plumage ('dirty' white). The **chick** has cream to fawn down, which is long and spiky on the head and darkest on the back.

The Brahminy Kite is a solitary or gregarious kite of inshore coastal and estuarine waters and adjacent terrestrial habitats, sometimes occurring over forest or inland along rivers in the tropics and subtropics. It has a rather loose rowing flight action, and the carpals are held forward when gliding, like the Whistling Kite. The Brahminy Kite glides on flat wings, which might be slightly raised when soaring (see figure 6.4). It glides or side-slips to snatch food from the tideline or water surface. Its voice is distinctive: a plaintive, descending cry, like that of a bleating lamb.

The juvenile Brahminy Kite can be confused with several species. The Osprey has a prominent head and bill, longer, narrower and more pointed wings, white belly, square tail, and heavy feet, and plunges into water. The Whistling Kite has pale upperwings, longer and more splayed primaries, bowed wings in glide, and a long tail. The pale Little Eagle is more robust with a pale upperwing band, crisper underwing pattern, square tail, and feathered tarsi. The immature Black-breasted Buzzard has upswept wings and a square tail.

Figure 6.4a Soaring and gliding

Figure 6.4b Gliding

DISTRIBUTION

The Brahminy Kite occurs around the northern Australian coast from about Shark Bay in the west to the Hunter River in the east, with occasional vagrants further south. It also occurs from India and South-East Asia to New Guinea and the Solomon Islands.

FOOD AND HUNTING

The Brahminy Kite eats a variety of small animals and carrion: mammals, birds, reptiles, amphibians, fish, arthropods, crustaceans, shellfish, cuttlefish, road-killed vertebrates, large carcasses, and offal. It forages by quartering and high soaring, or still-hunting from a prominent perch, and it sometimes searches on the ground. It seizes prey in a glide or dive that might become a short chase, hawks flying insects, snatches prey from the tree canopy and water surfaces, and robs other predators. It characteristically patrols water or the water's edge, and scoops food from the surface without submerging.

BEHAVIOUR

The Brahminy Kite is a conspicuous inhabitant of coastal areas including towns, where it can be seen soaring in the sea breeze, gliding along the shoreline, or scavenging around beaches and jetties. In display, males soar high over their territories, hang in the wind and side-slip across the wind, and perform long dives to spiral up again and repeat the process or glide to a lookout perch. Members of a courting pair circle together with calling, and sometimes perform long aerobatic dives interspersed with spiralling to regain height. They also perch prominently and call.

BREEDING

Laying takes place in the dry season in the tropics and late winter to spring in the subtropics. Pairs nest solitarily. The nest is a platform of sticks and other flotsam 40–60 cm across, up to 20 cm deep, lined with leaves and other soft material, including human rubbish, and placed 2–30 m above the ground in a tree or artificial structure. The clutch size is usually one or two eggs but ranges from one to four. Incubation takes about 35 days, and the nestling period is 50–56 days. Broods of one and two young, and occasionally three, fledge but usually only one. The period of dependence after fledging lasts up to two months.

THREATS AND CONSERVATION

The Brahminy Kite is not globally or nationally threatened. It is common to abundant and widespread in the tropics, where it benefits from extra food provided by humans in the form of carrion and refuse. Its range is contracting northwards in eastern Australia, where the thickness of its eggshells was reduced by DDT use and its population is affected by habitat disturbance.

White-bellied Sea-Eagle
Haliaeetus leucogaster

DESCRIPTION

This species' common name is a literal translation of the generic name ('sea-eagle') and specific name ('white belly'). The genus is characterised by powerful feet and bare tarsi with scutellate scalation in front and reticulate scaling behind, and rough spiny soles.

The White-bellied Sea-Eagle is 75–85 cm long, with a wingspan of 180–218 cm. The average weight of males is 2400 g, females 3330 g. It is similar in size to the Wedge-tailed Eagle but has shorter primaries and tail. It is a large soaring eagle with broad, upswept wings and a short, pale wedge-shaped tail. The adult is grey and white: unmistakable. The juvenile is mottled brown with pale bases to the primaries and a pale tail.

The **adult** is white with grey back, wings, and base of tail. The cere is pale grey to light blue-grey, the eyes brown, and the legs and feet cream. The **juvenile** is brown with pale spotting and scaling; the head is paler and mottled. The tail is white, grading to brown near the tip. The underwing has a pale panel across the bases of the primaries. The cere is grey to blue-grey, the eyes brown, and the legs and feet cream. **Immatures** (second to fourth or fifth year) become increasingly pale and similar to the adult; some at intermediate stages have a white head and underparts with a mottled breast band like that of the Osprey. The **chick** has white down.

The White-bellied Sea-Eagle is a solitary or loosely gregarious eagle of coasts, estuaries, rivers, inland lakes, and adjacent terrestrial habitats; it sometimes overflies other habitats. Its flies with a powerful rowing action; it glides and soars on broad, stiffly upswept wings with curved trailing edges (see figures 6.5 and 6.6). Juveniles have a more bulging trailing edge to the secondaries that appears serrated. At rest it is robust and chunky, with the wing tips enveloping or projecting slightly beyond the tail tip and the bare tarsi evident. Its voice is distinctive: a deep goose-like honking or cackling; begging juveniles give a more prolonged yelping or wailing.

The juvenile or immature White-bellied Sea-Eagle can be confused with the Osprey, but has shorter primaries, much broader, rounded, upswept wings, and a shorter, wedge-shaped tail. The Wedge-tailed Eagle is darker with longer wings and tail, feathered tarsi, and less upswept wings in gliding or soaring flight. The Black-

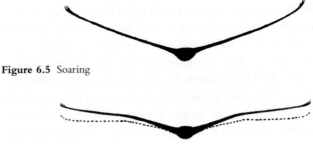

Figure 6.5 Soaring

Figure 6.6 Gliding

breasted Buzzard is darker (adult) or streaked black on the breast (immature), with more distinct panels in the wings, square tail, and more buoyant rocking flight.

DISTRIBUTION

The White-bellied Sea-Eagle occurs around the entire Australian coastline, including Tasmania, and extends well inland along rivers and around lakes and reservoirs. It also occurs from India and South-East Asia to New Guinea and the Bismarck Archipelago.

FOOD AND HUNTING

The White-bellied Sea-Eagle eats mammals, birds, reptiles, fish, carrion, and beach-cast offal and other scraps. Live prey includes rabbits, seabirds (as large as gulls, cormorants and gannets), spiny or poisonous fish, and sea-snakes. It forages by quartering, high soaring or still-hunting from a perch. It attacks in a shallow glide or dive to snatch prey from the ground or water surface (see figure 6.7); it harries aquatic birds to exhaustion, snatches fruit-bats from their tree roosts, robs other predators, and attends foraging dolphins. It scoops food from the water surface without submerging.

BEHAVIOUR

The White-bellied Sea-Eagle is a conspicuous and imposing sight around coasts and waterways, where it is commonly seen soaring gracefully or resting on a prominent perch. In the breeding season members of a pair engage in much mutual soaring, tandem flying, and mock attack-and-parry, with duet-calling in flight or on perches. Duetting and copulation by resident pairs also occur outside the breeding season. The adults defend nestlings aggressively against intruders, including humans.

BREEDING

The laying season is May to August in northern Australia and June to September, exceptionally November, in the south. Pairs nest solitarily. The nest is a pile of sticks 120–170 cm across, 50–180 cm deep, lined with leaves, grass, and seaweed, and placed on the ground or a cliff on offshore islands, otherwise 3–40 m above the ground in a tree. The clutch size usually is two eggs, ranging from one to three. Incubation takes about 40 days, and the nestling period is 65–70 days. Success has been measured as 1.3 young fledged per successful nest, 1.1 per active nest and 0.8 per occupied territory per year; 34% of successful nests raised two fledglings, 66% raised

Figure 6.7 Hunting

one. The period of dependence after fledging lasts two or three months, or perhaps longer. Young leave their natal territory about four months after fledging, and disperse up to 3000 km. Age at first breeding is 7 years old in captivity and not before acquisition of adult plumage (at about 5 years old) in the wild.

THREATS AND CONSERVATION

The White-bellied Sea-Eagle is not globally or nationally threatened. It is generally common, although there have been some localised declines in southern Australia through habitat destruction or disturbance to nest sites. It is scarce and localised in South Australia. It might have benefited from introduced prey species, and from rubbish dumps, in some areas. In Australia the thickness of its eggshells was reduced by DDT use. Shooting and other forms of direct persecution are a local problem. In South Australia, and probably elsewhere in southern Australia, the Sea-Eagle is at risk of further decline through increasing residential, tourist, and recreational developments.

Chapter 7

Harriers, genus *Circus*

The hawks in this cosmopolitan genus ('circling hawk') are so-called because of their low harrying flight back and forth across a patch of ground. They are characterised by a slim body, short broad head, long narrow wings, and long tail. The pattern of tarsal scalation is scutellate in front and reticulate behind. They fly buoyantly, gliding on raised wings low over open country or water. Their wings show five separated primary 'fingers', like those of the goshawks (*Accipiter*), rather than the six of large kites and small eagles. Harriers possess an owl-like facial ruff and large asymmetrical ear openings, apparently for detecting and pinpointing the sounds of prey in dense cover, and long legs for reaching into long grass or reeds. Depending heavily on sound for hunting, they have aural acuity better than that of most raptors and approaching that of the owls.

Harriers constitute the only group of raptors in which polygyny is common, although in Australasia the Swamp Harrier is usually monogamous and the Spotted Harrier invariably so. The polygynous harriers are sexually dichromatic, with distinctively coloured males and cryptically coloured (juvenile-like) females. Harriers generally nest on the ground or among wetland vegetation. The Spotted Harrier nests in trees. It builds a typical harrier 'ground nest' in character, although in precarious positions on the horizontal limbs or forks of trees. Tree-nesting in the Spotted Harrier is probably an adaptation to avoid ground predators in the dry country where it lives.

Most harriers fit into one of two recognisable groups: the small dry-country harriers of the Americas and Old World, and the large marsh-harriers of the Old World and Australasia. David Baker-Gabb has proposed a southern (Gondwanan) origin for the genus. In support of this theory is the apparently ancestral position of the Spotted Harrier and the similarity of its plumage to that of the Cinereous Harrier (*Circus cinereus*) of South America and the similarity of its aerial display to that of the African Marsh-Harrier (*Circus ranivorus*). Furthermore, the Swamp Harrier seems more closely related to the African Marsh-Harrier than to the Northern or Eastern Marsh-Harriers (*Circus aeruginosus* and *Circus spilonotus*).

Spotted Harrier
Circus assimilis

DESCRIPTION

The specific name ('similar') indicates a supposed likeness to others in the genus, although the Spotted Harrier is distinctive in plumage. The Spotted Harrier is 50–61 cm long (tail about half), with a wingspan of 121–147 cm. The average weight of males is 465 g, females 671 g. It is similar in size to the Whistling Kite and Swamp Harrier, with a lankier build than the latter. The adult is unmistakable, with grey upperparts and chestnut underparts spotted with white. The juvenile is ginger and brown, with pale upperwing coverts, heavily barred tail, and fawn rump.

The **adult** has slate-grey upperparts with chestnut, grey-streaked head and face, chestnut shoulder patch, white spots on the wings, and dark bars on the secondary and tail feathers. The underparts are chestnut spotted with white, grading to fine barring on the thighs. The undersides of the remiges are pale and narrowly barred, with dark trailing edges and outer primaries. The cere is yellow (occasionally paler), the eyes are yellow to orange-yellow, and the legs and feet are yellow. The **juvenile** has dark brown upperparts with buff feather tips, buff patch on the forewings, fawn rump, and broad dark bars on the tail. The head and face are pale rufous, streaked brown. The underparts are buff with fine dark streaks. The underwings are pale and narrowly barred on the remiges, with dark tips and trailing edges; the undertail is thickly barred. The cere is pale yellow (sometimes tinged green), the eyes dark brown, and the legs and feet yellow. **Immatures** pass through a stage (second year) when the plumage is like that of an adult but more mottled above and streaked below; the streaking finishes abruptly on the chest in some birds (giving a dark-hooded appearance like the male Papuan Harrier). Immatures then go through another stage (third year) when the plumage is like that of an adult but with white streaks, not spots, on the rufous underparts. The **chick** has grey first down with a white facial ruff and grey-brown second down.

The Spotted Harrier is a solitary harrier of crops, grassland, low shrubland, and open woodland in inland and northern Australia, although it sometimes occurs over coastal grassland, heath or swamps in the south. Its flight is buoyant with gentle, rhythmic wing-beats and extended glides. It sails low to the ground with wings held in a dihedral (see figure 7.1); the tail has a wedge-shaped tip that looks 'dished' from behind (edges curved up). The wings are narrowest at the base, broadening to deeply 'fingered' tips. It sometimes soars high in circles, and the legs might be lowered during low sailing or high soaring flight. It perches on the ground or in trees. At rest it sits high on long legs, with the tail tip projecting just beyond the wing tips. It is usually silent, but breeding birds utter piercing squeaks and a rapid chatter.

The juvenile Spotted Harrier can be confused with the adult Swamp Harrier, which has a heavier build, white (not fawn) rump, and lightly barred tail with a gently rounded to almost square tip, and lacks a pale upperwing patch. The Square-tailed Kite has longer

Figure 7.1a Soaring and gliding

Figure 7.1b Gliding

barred primaries, a square or notched tail twisted in flight, and short legs. The Red Goshawk is more robust with a crested nape, quicker flight, less upswept wings in glide, square tail, and heavy legs and feet.

DISTRIBUTION

The Spotted Harrier occurs throughout Australia but not in densely forested parts, and it rarely reaches Tasmania. It also occurs in Sulawesi and the Lesser Sunda Islands.

FOOD AND HUNTING

The Spotted Harrier eats terrestrial birds (quails, larks, pipits), mammals (young rabbits, rodents), reptiles, large insects, and (rarely) carrion. It forages by low, slow quartering or transect hunting, sometimes by hovering. It seizes prey by diving to the ground or by a short chase if prey is flushed.

BEHAVIOUR

The Spotted Harrier is usually seen sailing languidly, at or near ground level. It is more silent and less inclined to perform aerial displays than is the Swamp Harrier, but does spiral high and occasionally performs diving displays. Its aerial Undulating Display, although rarely seen, is similar to that of the Swamp Harrier and the African Marsh-Harrier. In courtship, the male escorts the female to the nest where he performs a Bowing Display (see figures 7.2 and 7.3). Also in courtship, and when provisioning the incubating or brooding female, the male delivers prey in an aerial food pass (see figure 7.4). In southern Australia Spotted Harriers are spring–summer breeding migrants, generally moving northwards for the winter, but the extent, timing, and location of migratory movements vary with seasonal conditions and the availability of food.

BREEDING

The laying season is usually September to October in southern Australia, exceptionally March to June, and June to September in the north. There might be two breeding seasons in a year. Pairs nest solitarily. The nest is a platform of sticks lined with green leaves,

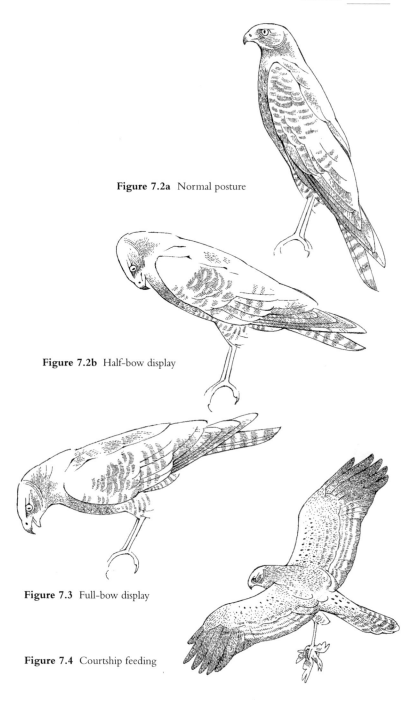

Figure 7.2a Normal posture

Figure 7.2b Half-bow display

Figure 7.3 Full-bow display

Figure 7.4 Courtship feeding

40–70 cm across, 16–30 cm deep, and placed 2–15 m above the ground in a living tree. There are rare cases of unsuccessful nests built on the ground. The clutch size is usually three eggs, ranging from two to four. Incubation takes 33 days, and the nestling period is 36–43 days. Success has been measured as 56–75% hatching success and 44–57% fledging success (of eggs laid), and as 1.3 young fledged per nest, 2.17 per successful nest, and 0.9 per territorial pair. The period of dependence after fledging lasts at least six weeks, after which juveniles disperse or migrate up to 1600 km. Age at first breeding is two years.

THREATS AND CONSERVATION

The Spotted Harrier is not globally or nationally threatened. It is generally uncommon but widespread; it might have benefited locally in southern Australia from the creation of suitable habitat and an increase in native and introduced prey. The thickness of its eggshells was not significantly reduced by DDT use in Australia.

Swamp Harrier
Circus approximans

DESCRIPTION

The specific name ('approaching') alludes to the similarity of this species to other harriers, and probably the Northern Marsh-Harrier specifically.

The Swamp Harrier is 50–61 cm long (tail almost half), with a wingspan of 121–142 cm. The average weight of males is 647 g, females 890 g. It is similar in size to the large kites and Spotted Harrier, with a somewhat heavier build than the latter. It is distinguished by a white or pale rump.

The **adult male** has a tricoloured pattern above, of brown saddle, grey wings, and black wing tips; it has a white rump and a grey, lightly barred tail. The underparts are white streaked with rufous. The underwings and undertail are faintly barred. The cere is yellow, the eyes pale yellow, and the legs and feet yellow to orange-yellow. The **adult female** is darker and browner than the male, with a grey wash to the remiges; the underparts are buff heavily streaked rufous-brown. The cere is yellow, the eyes light brown to pale yellow (older birds), and the legs and feet yellow. The **juvenile** has dark brown upperparts with white streaks on the nape, a rufous rump, and faint bars on the tail, which is washed orange. The underparts are dark brown, fading to rufous. The underwings are unbarred, with a pale patch towards the tips. The cere is yellow, the eyes dark brown (changing to light brown with age in males), and the legs and feet yellow. The **immature male** is similar to the adult male, but the eyes are golden (second year) to pale yellow (third year). The **immature female** is similar to the adult female, but the underparts are darker brown and the eyes are brown. The **chick** has fawn to white first down with a white 'skull-cap' and facial mask, and grey to buffy white second down.

The Swamp Harrier is a solitary or loosely gregarious harrier of lakes, swamps, grassland, coastal heath, and tall crops. Its flight action is buoyant, with smooth, rhythmic wing-beats and long glides, holding the wings in a dihedral, and with slight rocking or tilting from side to side (see figure 7.5); the outerwings are slightly narrower than the innerwings, with the female having a more curved trailing edge to the wing than the male. It soars in circles and sometimes lowers the legs in flight. It perches on the ground or on low posts, stumps or swamp vegetation. At rest it sits high on long legs, with the wing tips reaching, or almost reaching, the tail tip. It is usually silent, but soaring birds sometimes utter a high-pitched, descending mew.

Figure 7.5a Soaring and slow gliding

Figure 7.5b Fast gliding

The adult Swamp Harrier can be confused with the juvenile Spotted Harrier, which has a more ginger tone, pale forewing patch, fawn (not white) rump, heavily barred tail with a wedge-shaped tip, and more buoyant flight. The Square-tailed Kite has longer barred primaries, a square or notched tail that is twisted in flight, and short legs. The Brown Falcon is smaller, with prominent a double cheek mark, more pointed and less upswept wings, and lacks a pale rump. Swamp Harriers with rufous underparts and barred underwings and tail are often mistaken for the Red Goshawk, which is more robust with a crested nape, flies more rapidly, has less upswept wings when gliding, and heavier legs and feet, and lacks the white rump. The Swamp Harrier is difficult to distinguish from the female and juvenile Papuan Harrier, which might rarely occur in northern Australia (for differences, see the account for that species).

DISTRIBUTION

The Swamp Harrier occurs through most of Australia and Tasmania, except densely forested parts and the arid inland away from surface water. Breeding has been recorded mainly in southern regions, although its occurrence has recently been confirmed in coastal Queensland as far north as the Townsville region. The Harrier also occurs in New Zealand and Polynesia. It is a winter migrant to New Guinea and is resident elsewhere in Melanesia, for example, Fiji. It is a rare vagrant to subantarctic islands.

FOOD AND HUNTING

The Swamp Harrier eats mammals, birds and their eggs, reptiles, amphibians, fish, insects, and carrion. It forages by low, slow quartering as well as by soaring. It seizes prey by diving or dropping to the ground or water surface, sometimes after hovering; it harries waterbirds to exhaustion or sometimes drowns them. It also robs other raptors.

BEHAVIOUR

The Swamp Harrier is usually seen sailing low over wetlands or open country. It often perches or roosts on the ground. In the breeding season, birds soar high and perform a languid Undulating Display with deep, sweeping wing-beats and some twisting and rolling at the zenith, accompanied by mewing calls. The male also dives at the female, who either evades him or rolls to parry with her feet. The female's descent to the nest is followed by the male descending in stages like a falling leaf. Food passes, in courtship and to provision the nesting female, are aerial. Males mark their territorial boundaries by flying along them with the wings held high and the legs down (see figure 7.6a and b). After breeding, Swamp Harriers in southern regions (notably Tasmania) migrate northwards, sometimes in groups. Wintering birds roost communally; such roosts might function as 'information centres' at which some birds learn where others are hunting successfully.

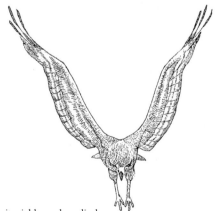

Figure 7.6a Territorial boundary display

Figure 7.6b Territorial boundary display

BREEDING

The laying season is September to December. Pairs nest solitarily or in loose 'clumps' with other pairs; the species is rarely polygynous. The nest is a platform of sticks, reeds, grass, and other plants among tall grass, shrubs or reeds, either on the ground or in water. The clutch size is usually three or four eggs, although ranging from two to seven. Incubation takes 33 days, and the nestling period is 43–46 days. Success has been measured as 1.05–1.75 young fledged per nest and 1.82–2.15 per successful nest. The period of dependence after fledging lasts about four to six weeks, after which juveniles disperse at about seven weeks and wander or migrate up to 1500 km.

THREATS AND CONSERVATION

The Swamp Harrier is not globally or nationally threatened. It is common in suitable habitat, but has declined where wetlands have been drained. Nests are very vulnerable to human disturbance. The thickness of its eggshells was reduced in agricultural areas of southern Australia by the use of DDT.

Swamp Harrier *Circus approximans* **1** Adult male, pale; **2** Adult male, dark; **3** Adult female;
4 Juvenile; Spotted Harrier *Circus assimilis* **5** Adult female; **6** Juvenile

Spotted Harrier *Circus assimilis* **1** Adult female; **2** Juvenile, fresh plumage; **3** 'Dark-hooded' first immature (in post-juvenile moult); **4** First immature; **5** Second immature; Swamp Harrier *Circus approximans* **6** Adult male, pale; **7** Adult male, dark; **8** Adult female; **9** Juvenile

Spotted Harrier *Circus assimilis* **1** Adult female; **2** Juvenile, fresh plumage; **3** 'Dark-hooded' first immature (in post-juvenile moult); **4** First immature; Swamp Harrier *Circus approximans* **5** Adult male, pale; **6** Adult male, dark; **7** Adult female; **8** Juvenile

Brown Goshawk *Accipiter fasciatus,* Nominate *fasciatus:* **1** Adult female; **2** Juvenile male;
3 Immature male; Subspecies *didimus:* **4** Adult female; Subspecies *natalis:* **5** Adult female;
6 Juvenile male; Collared Sparrowhawk *Accipiter cirrhocephalus* **7** Adult male, pale; **8** Adult
female, dark; **9** Juvenile male

A female Brown Goshawk attacking prey *(Nicholas Birks)*

A juvenile female Collared Sparrowhawk resting at midday *(Nicholas Birks)*

Brown Goshawk *Accipiter fasciatus* **1** Adult female, nominate *fasciatus;* **2** Juvenile male, nominate *fasciatus;* **3** Immature male, nominate *fasciatus;* **4** Adult female, subspecies *didimus;* **5** Adult female, subspecies *natalis;* **6** Juvenile male, subspecies *natalis;* Collared Sparrowhawk *Accipiter cirrhocephalus* **7** Adult female; **8** Juvenile male

Peregrine Falcon *Falco peregrinus* (subspecies *macropus*) **1** Adult; **2** Juvenile; Grey Falcon *Falco hypoleucos* **3** Adult; **4** Juvenile; Australian Hobby *Falco longipennis* **5** Adult, dark; **6** Juvenile, dark*;* **7** Adult, pale; Nankeen Kestrel *Falco cenchroides* (nominate *cenchroides*); **8** Adult male; **9** Adult female; **10** Juvenile; **11** Immature male

Brown Goshawk *Accipiter fasciatus* **1** Adult female, nominate *fasciatus;* **2** Juvenile male, nominate *fasciatus;* **3** Immature male, nominate *fasciatus;* **4** Adult female, subspecies *didimus;* **5** Adult female, subspecies *natalis;* **6** Juvenile male, subspecies *natalis;* Collared Sparrowhawk *Accipiter cirrhocephalus* **7** Adult female; **8** Juvenile male

Eastern Marsh-Harrier
Circus spilonotus

DESCRIPTION

Black-and-white harriers, observed on rare occasions in the Australian tropics, are probably referable to the endemic New Guinean race (*Circus spilonotus spilothorax*) of this species ('spotted back'), although this has yet to be confirmed, and such birds need to be distinguished carefully from others such as the Asian Pied Harrier (*Circus melanoleucos*). The race concerned is commonly given its own vernacular name of Papuan Harrier. At present, the occasional black-and-white adult or near-adult males seen are assumed to be vagrants. However, it has been claimed that a pair of this species bred in Australia (a pair with flying young at the tip of Cape York). Females and juveniles might go undetected among Swamp Harriers in the tropics.

The Papuan Harrier is 48–53 cm long (tail about half), with a wingspan of about 110–130 cm. It is slightly smaller than the Swamp Harrier. The adult male is strikingly pied; the female and juvenile are similar to the Swamp Harrier in equivalent plumages.

The **adult male** has black upperparts, with much grey in the wings, a white rump, and a grey tail. The underparts are white with a black throat and variably black-streaked breast (streaks sometimes absent or nearly so). The underwings are white with black tips and trailing edges. The cere is greenish yellow, the eyes pale yellow, and the legs and feet orange-yellow. A melanistic morph is mostly black with a white rump and grey tail. The **adult female** has brown upperparts with a pale rump and heavily barred tail; some birds have white streaks on the head. The underparts are streaked rufous-brown; the underwings are barred with dark trailing edges; and the undertail is strongly barred. The eyes, legs, and feet are yellow. A melanistic morph is mostly dark brown, paler on the head, with pale buff patches on the nape, scapulars and rump; the tail is light brown with dark bars, and the eyes are dark. The **juvenile** is dark brown with variable cream streaks on the face, head, and nape; buff streaks on the back; and a faintly barred tail. The underwings are barred with a pale patch towards the tips. The eyes are brown. The **immature male** is more grey and white than the adult, with variable white mottling on the head and upperparts, dark streaks on the underparts, and spots on the underwing coverts.

The male Papuan Harrier is similar to the male Pied Harrier of Asia, which is much smaller and has more white on the leading edges of the innerwings. The female is similar to the Swamp Harrier but has dark trailing edges to the underwings and might have white head markings. The juvenile is similar to the juvenile Swamp Harrier but has pale streaks on the head and back. 'Dark-hooded' immature Spotted Harriers, before completion of their post-juvenile moult, can be mistaken for male Papuan Harriers in some views.

DISTRIBUTION

Black-and-white harriers have been seen in far northern Queensland, the Top End of the Northern Territory, and the Kimberley region of Western Australia, and on islands in Torres Strait. They are most likely the Papuan Harrier, a subspecies of the Eastern Marsh-Harrier, from nearby New Guinea. However, other black-and-white harriers (Pied Harrier and Asian form of the Eastern Marsh-Harrier, nominate race *spilonotus*) migrate to South-East Asia from more northerly latitudes; they could occasionally 'overshoot' beyond the Equator and perhaps reach Australia.

Chapter 8

Goshawks and sparrowhawks, genus *Accipiter*

This cosmopolitan genus (*Accipiter* means 'hawk'), of about forty species, is the largest and most diverse in the family. Males of the smallest species, at under 100 g, are thrush-sized, whereas females of the largest species, at more than 1000 g, are as big as small eagles. Collectively, they are known elsewhere as 'wood-hawks' because they are adapted for swift, agile flight within forest and woodland. All are characterised by broad rounded wings (falling well short of the tail tip at rest), long tails, and long legs and toes. They also have tight plumage, which, in adults, is usually finely barred below. The larger species have a fierce expression, caused by heavy brow ridges. The tarsi have scutellate scalation in front and reticulate scaling behind. This pattern is obvious in juveniles, but the scutes tend to fuse to produce smooth tarsi in adults.

These hawks hunt from perches or by flying stealthily and dextrously among trees. They have perfected the technique of ambush waiting, suddenly bursting from cover and accelerating rapidly in a short chase if necessary. The sparrowhawks have spindly legs and long toes, notably the middle toe, for snatching birds in flight; the larger goshawks are more heavily 'armed' and sometimes plunge recklessly into cover after their quarry, which might be as large as themselves. The name *goshawk* (pronounced gos-hawk) shares the same derivation as *gosling*, and refers to the medieval practice of using hawks to hunt geese and other game for the pot.

The distinction between goshawk and sparrowhawk, based on size and foot structure, is rather arbitrary and does not necessarily reflect genetic relationships. The Brown and Grey Goshawks, together with several New Guinean and Melanesian species, form a related group derived from an endemic radiation. The Collared

Sparrowhawk is related to several New Guinean and Wallacean sparrowhawks; as a group, they are probably related to the Old World sparrowhawks.

A New Guinean species, the Grey-headed Goshawk (*Accipiter poliocephalus*, the specific name meaning 'grey head'), might yet be confirmed as occurring in tropical rainforest on Cape York Peninsula. Unsubstantiated claims have culminated in an unconfirmed report of a pair nesting at the tip of Cape York Peninsula, a sighting on Saibai Island in Torres Strait, and a sighting in far northern Queensland of an unidentified goshawk that might have been a juvenile of this species. The Grey-headed is like a small, unbarred Grey Goshawk (grey morph) with more orange-red cere, eye-rings, and legs. Claims of the Grey-headed Goshawk in Australia require careful substantiation (see New Guinea references in the bibliography).

Brown Goshawk
Accipiter fasciatus

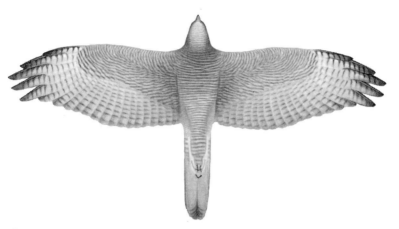

DESCRIPTION

The specific name ('banded') refers to the ventral barring. The species is characterised by prominent bony brow ridges and robust feet. The tip of the middle toe does not project beyond the claws of the other toes.

Southern Brown Goshawks (race *fasciatus*) are 40–55 cm long (tail about half), with a wingspan of 74–96 cm. The average weight of males is 350 g, females 570 g (Tasmanian females are heavier: 614 g). Tropical birds (race *didimus*) are smaller: 37–46 cm long, with a wingspan of 70–85 cm, and males weigh 227 g and females 343 g. The Brown Goshawk is slightly smaller than the Brown Falcon, with shorter, broader wings. It is a fierce, active hawk with rounded wings, long rounded tail and long legs.

The **adult** of the nominate race (*fasciatus*) has slate-grey upper-parts, sometimes washed brown (especially in females), and a rufous half-collar. The underparts are finely barred dull rufous and white, and the underwings and tail are finely barred. The cere is cream to olive-yellow, the eyes yellow, and the legs and feet yellow to orange-yellow. The **juvenile** has brown upperparts with pale streaks on the head and nape, and fine rufous edges to the feathers of the back and wings. The underparts are white with heavy brown streaks on the breast and coarse brown barring on the belly. The underwings and tail are finely barred. The cere is olive-yellow, the eyes brown (in fledglings) to pale yellow, and the legs and feet pale yellow. The **immature** (second year) has brown upperparts without pale streaks or rufous edges to feathers; some have a chestnut half-collar. The underparts are barred brown and white, somewhat more coarsely than in the adult. The adult of the northern race (*didimus*), confined to tropical coastal and subcoastal regions, is 10% smaller than the nominate race, with paler grey upperparts and more rufous underparts in adult plumage, and slightly more pointed wing tips in flight. The **chick** has white first down and light salmon-brown second down with a white 'skull cap'.

The Brown Goshawk is a solitary, secretive hawk of most wooded habitats, including farmland and trees in urban areas. Its flight action is rapid and powerful, with bursts of quick, deep wing-beats; it glides on slightly bowed wings and soars on slightly upswept wings (see figures 8.1 and 8.2). At rest it sits high on long legs, with the short wings not reaching the tail tip. It gives a variety of rapid and sometimes shrill or slower and mellow chattering calls as well as single upslurred mewing notes repeated at about one note per second.

The Brown Goshawk's plumage is similar to that of the Collared Sparrowhawk. In flight the former is best distinguished by its longer head and neck, straighter trailing edges to the wings, longer rounded tail and heavier, deeper wing-beats (see figures 8.3 and 8.4).

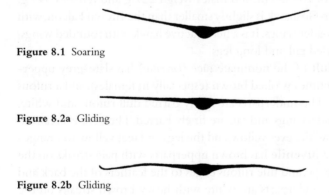

Figure 8.1 Soaring

Figure 8.2a Gliding

Figure 8.2b Gliding

Figure 8.3 Soaring (a) Brown Goshawk (b) Collared Sparrowhawk

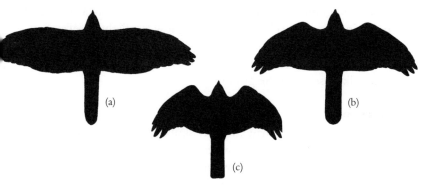

Figure 8.4 Gliding (a) Brown Goshawk, slow glide (b) Brown Goshawk, fast glide (c) Collared Sparrowhawk

At rest it has a more beetle-browed, menacing expression, sturdier legs, and heavier feet with a shorter middle toe. The dorsal colour of some Brown Goshawks is like that of the Grey Goshawk, which has broader, more rounded wings and a shorter, squarer tail. A distant Pacific Baza is similar in colour but is more boldly barred, with long splayed primaries and slow flight; at rest it shows a crest, long wings almost reaching the tail tip, and short legs. A perched Brown Falcon has dark facial markings, dark eyes, and long wings reaching the tail tip.

DISTRIBUTION

The Brown Goshawk occurs throughout mainland Australia and Tasmania, and Wallacea, New Guinea, New Caledonia and the south-easternmost Solomon Islands. The form inhabiting Christmas

Island, in the Indian Ocean off Java, has traditionally been assigned to this species, but might belong to another species (see the Grey Goshawk account).

FOOD AND HUNTING

The Brown Goshawk eats mammals, birds, reptiles, amphibians and arthropods, and occasionally carrion; it preys mostly on birds and young rabbits in the south, and birds and lizards in the north. Vertebrate prey usually weighs less than 500 g, but sometimes more than a kilogram. It forages mostly by still-hunting from a concealed perch in foliage as well as by soaring or low fast flight. It seizes prey, usually on the ground, by a stealthy glide or direct flying attack that sometimes becomes a short chase. It also flushes prey from cover and stalks insects on the ground.

BEHAVIOUR

The wary Brown Goshawk usually skulks, unobserved, in cover so that it is often only glimpsed when it flushes and leaves through the trees. It is more readily seen when it soars in spirals. It can also be seen at close quarters when it is engrossed in the hunt. In urban or rural situations, this might be when it recklessly attacks domestic birds or chickens around houses. In the breeding season, the Goshawks perch and call conspicuously and perform aerial displays. Single birds soar and perform an Undulating Display; pairs soar and perform chasing or following flights with deep, exaggerated beats (Slow-flapping Flight) and upslurred, ringing whistles.

Brown Goshawks of the southern race are partly migratory over most of their range: juveniles and some adults, particularly females, from high altitudes and high latitudes winter in coastal, lowland, or northern parts of Australia, exceptionally reaching islands to the north. Some adults defend regular winter territories separate from their breeding territories. There is a noticeable influx of mostly young birds into urban areas from early autumn. Some individuals of the tropical race, from inland northern Australia, winter on the northern coast and perhaps in parts of Wallacea. An unresolved question is whether the two races interbreed or behave as separate species in areas where their ranges overlap.

BREEDING

The laying season is September to December. Pairs breed solitarily. The nest is a platform of sticks 38–70 cm across, 18–30 cm deep, lined with green leaves, and placed 2–36 m above ground in the fork of a living tree. The clutch size is usually three eggs but ranges from two to four. Incubation takes 30 days, and the nestling period is 28–37 days. Success was 1.6–1.75 young per attempt and 2.1–2.75 per successful nest in south-eastern Australia; 62% of attempts were successful in north-eastern Australia, with 1.6 young per successful nest and 1.0 young per active nest. The period of dependence after fledging lasts up to six weeks, after which juveniles disperse or migrate up to 900 km. Sexual maturity is attained at one year, although breeding in juvenile plumage is rare. Mean annual survival is 65–79% in south-eastern Australia, with banded birds living up to 17 years.

THREATS AND CONSERVATION

The Brown Goshawk is not globally or nationally threatened. It is common and widespread. There have been local declines in southern Australia where habitat clearance is extensive, but it has benefited from the introduction of the rabbit, and it preys on introduced birds. The effects of DDT on its eggshell thickness were local and insignificant in Australia. It is often illegally shot (mainly juveniles), but the effect on populations is also insignificant. The isolated form on Christmas Island is threatened and classed as Vulnerable, owing to its small range and population size (50–150 pairs), past habitat clearance, and persecution, which has largely stopped. Its biology is little known, and it might belong to a different species (see the Grey Goshawk account).

Grey Goshawk
Accipiter novaehollandiae

DESCRIPTION

The specific name ('of New Holland') denotes the country of origin of the first specimen known to science. Internationally it is now called the 'Variable Goshawk' or, perhaps better, the 'Varied Goshawk', but this might be premature pending resolution of the taxonomic status of the populations on tropical islands.

The Grey Goshawk is 38–55 cm long (tail slightly less than half), with a wingspan of 71–110 cm. The average weight of males is 356 g and females 720 g (although in Tasmania birds are heavier: males 415 g, females 797 g). It is slightly larger and more heavily built than the Brown Goshawk, with broader wings and a shorter, squarer tail. It is a striking grey-and-white or all-white goshawk with bright orange-yellow cere and feet. The wings are very broad and rounded, and the legs and feet are massive.

Plumage dimorphism occurs in both sexes; only the white morph occurs in Tasmania and it predominates in Victoria and Western Australia. The **adult grey morph** has grey upperparts and white underparts with fine grey barring on the breast, and faintly barred underwings and tail. The cere is orange-yellow, the eyes deep red, and the legs and feet orange-yellow. The **juvenile grey morph** is similar to the adult, with coarser, wavy bars on the breast and a slight brown collar. The cere is yellow, the eyes brown (in fledglings) to yellowish brown or orange (second year). The **adult white morph** is entirely white, and the bare parts are as for the adult grey morph. The **juvenile white morph** is entirely white, although some birds have faint barring on the underwings and tail; the bare parts are as for the juvenile grey morph. The **chick** has white down.

The Grey Goshawk is a solitary, secretive hawk of tall, wet forest in eastern and south-eastern Australia and riverine forest in northern Australia. It occasionally appears in more open woodland and urban areas. Its flight action is quick but sometimes appears shallower and more laboured than that of the Brown Goshawk. The Grey Goshawk soars and glides on bowed wings, or slightly upswept wings when soaring to gain height rapidly (see figures 8.5 and 8.6). The tail is sometimes widely fanned. At rest it sits high on long legs, with the long tail extending well beyond the wing tips. Its most commonly heard call is a series of mellow, rather nasal, upslurred ringing whistles (one note per second).

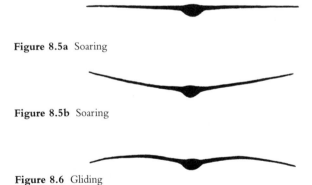

Figure 8.5a Soaring

Figure 8.5b Soaring

Figure 8.6 Gliding

Pale-backed adult Brown Goshawks can be mistaken for the grey morph but have narrower wings, a longer, more rounded tail, and coloured underparts. The Grey Falcon has long, narrow, pointed wings with black tips (extending to the tail tip at rest), shorter legs, and winnowing or flickering flight. The *Elanus* kites have black forewings, black marks on the underwings, long pointed wings, short white tails, short legs, and hovering flight. The Pacific Baza has longer wings with splayed, barred primaries, slow flight, boldly barred belly, and short legs. The white morph is distinguished from white cockatoos by its more powerful flight, longer tail, long legs, and lack of coloured crest or underwings.

DISTRIBUTION

The Australian form, a large form with only grey and white morphs, occurs in northern, eastern, and south-eastern Australia, including Tasmania. Small, brightly coloured polymorphic forms occur in Wallacea, New Guinea, and the Solomon Islands. However, these latter forms might constitute a separate species, *Accipiter hiogaster*, which would be known as the Varied Goshawk. The form on Christmas Island, usually regarded as a race of the Brown Goshawk, might belong to the island Varied Goshawk complex.

FOOD AND HUNTING

The Grey Goshawk eats mammals, birds, reptiles, amphibians, arthropods, and (rarely) carrion. Prey ranges up to rabbit and heron size in southern Australia. It forages mainly by still-hunting from a concealed perch in the tree canopy as well as by low fast flight, quartering and soaring. It seizes prey on the ground or a perch by diving, stealthy glides or direct flying attacks that become a short chase. It also drags ringtail possums (*Pseudocheirus* species) from their dreys.

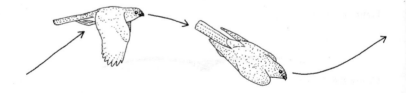

Figure 8.7 Mutual flying displays

BEHAVIOUR

Both morphs of the Grey Goshawk are cryptic and difficult to see when perched high within their preferred habitat of rainforest or giant, pale-trunked forest gums, although their presence is betrayed by their calls. They also perch for long periods in exposed positions, high in tall trees on creeklines or at the edges of forests. They are more readily seen when they soar in spirals above the forest. At such times, in the breeding season, they also perform solitary Undulating Displays and paired Slow-flapping Flights like those of the Brown Goshawk (see figure 8.7). In the non-breeding season, individuals can be seen when they boldly attempt to take domestic chickens or birds around urban or rural houses.

BREEDING

The laying season is September to December in southern Australia and May to November in the north. Pairs nest solitarily. The nest is a platform of sticks 50–60 cm across, 35 cm deep, lined with green leaves, and placed up to 15 m above the ground in the canopy of a living tree. The clutch size is usually two or three eggs but is

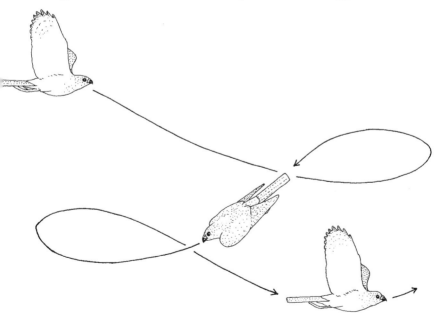

occasionally four. Incubation takes about 30 days, and the nestling period is 30–38 days. Nine young fledged from 14 eggs; 94% of attempts were successful, with 1.5 young per successful nest and 1.4 young per active nest. The period of dependence after fledging lasts up to six weeks, after which young disperse widely. Longevity in captivity is up to 21 years.

THREATS AND CONSERVATION

The Grey Goshawk is not globally or nationally threatened. It is common in the tropics and subtropics but uncommon at the extremities of its range in southern and north-western Australia. It is subject to habitat loss in south-eastern Australia. The thickness of its eggshells was not significantly reduced by the use of DDT in Australia.

Collared Sparrowhawk
Accipiter cirrhocephalus

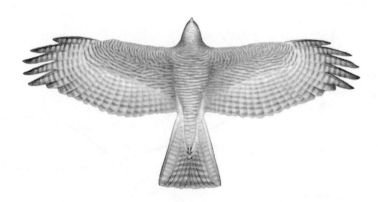

DESCRIPTION

The specific name ('tawny head') refers to the rufous half-collar around the hindneck. The species is characterised by slight brow ridges and slender feet. The last segment of the middle toe projects beyond the claws of the other toes.

The Collared Sparrowhawk is 29–38 cm long (tail about half), with a wingspan of 55–78 cm. The average male weighs 126 g, females 218 g. The male is similar in size to the Nankeen Kestrel but has shorter, broader wings; the female is similar in size to the male Brown Goshawk, although slimmer. It is a small, fierce, finely built hawk with rounded wings, long square-tipped tail, staring yellow eyes, and long legs.

The **adult** has slate-grey upperparts, sometimes with a brown wash, and a chestnut half-collar. The underparts are finely barred rufous and white. The underwings and tail are finely barred. The cere is cream to olive-yellow, the eyes yellow, and the legs and feet yellow. The **juvenile** has brown upperparts, with pale streaks on the head and nape, and fine rufous edges to the feathers of the back and wings. The underparts are white with heavy brown streaks on the breast and coarse brown barring on the belly. The underwings and tail are finely barred. The cere is cream to greenish yellow, the eyes brown (in fledglings) to pale yellow, and the legs and feet pale yellow. The **chick** has white first down and greyer second down with a white 'skull cap'.

The Collared Sparrowhawk is a solitary, secretive hawk of most well-wooded habitats, including farmland, urban areas, and drainage lines in the inland. Its flight is usually rapid with bursts of quick wing-beats; low direct flight is variously flickering and jerky ('hedge-hopping' hunting flight), undulating like that of a cuckoo-shrike, or occasionally slow with wing-beats like that of a Pacific Baza. It glides on flat or slightly drooped wings, and soars on slightly upswept wings (see figures 8.8 and 8.9). At rest it sits high on long legs, with the short wings not reaching the tail tip. Some calls are more rapid, thinner, and higher-pitched than those of the Brown Goshawk. Its most commonly heard call is a very rapid chitter like the alarm call of the White-plumed Honeyeater (*Lichenostomus penicillatus*); also a series of peevish notes like the Sacred Kingfisher (*Todiramphus sanctus*) but more slurred, sometimes breaking into the chitter. Begging juveniles give a distinctive loud, slow, downslurred mew repeated in series (one note per second).

The Collared Sparrowhawk's plumage is similar to that of the Brown Goshawk. In flight, the Sparrowhawk is best distinguished

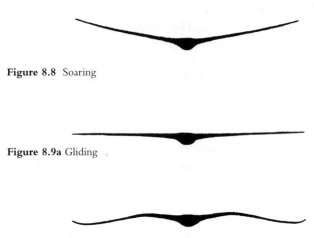

Figure 8.8 Soaring

Figure 8.9a Gliding

Figure 8.9b Gliding

by its silhouette, with smaller head, more curved trailing edges to the wings, and square or notched, sharp-cornered tail tip, and by its more winnowing or flickering, jerky flight (see figures 8.3 and 8.4). At rest, the Sparrowhawk has more staring eyes (lacking the Brown Goshawk's heavy brow-ridges), more spindly legs, and longer middle toe. The Australian Hobby, a fast-flying falcon, has a dark head pattern, dark eyes, long, narrow pointed wings (reaching the tail tip at rest), and short legs. The Pacific Baza has bolder barring, long splayed primaries, and slow flight; at rest it shows a crest, long wings almost reaching the tail tip, and short legs.

DISTRIBUTION
The Collared Sparrowhawk occurs throughout mainland Australia and Tasmania as well as in New Guinea and some of its satellite islands.

FOOD AND HUNTING
The Collared Sparrowhawk mostly eats small birds, particularly passerines (including introduced sparrows and starlings); it also takes lizards, insects, and (rarely) small mammals. Most prey weighs less

than 100 g and rarely more than 200 g. It forages by short-stay perch hunting from a concealed position in foliage, punctuated by short tree-to-tree, often undulating, flights. It also forages by quartering flight or low fast flight, sometimes hedge-hopping. Prey is seized in flight by a direct flying attack or a stealthy glide that becomes a short chase.

BEHAVIOUR

The Collared Sparrowhawk is often trusting and approachable, although cryptic and therefore easily overlooked. Although difficult to see when perched in cover, its presence is betrayed by its calls. It is more readily seen in soaring flight. Early in the breeding season, solitary birds or pairs soar and perform Undulating Displays and Slow-flapping Flight with chittering calls. Individuals also perch in the nesting area and give slower, mellow chattering calls. The male gives soft mewing notes as he brings food to the female, and mating is accompanied by loud squealing calls. Later in the breeding cycle, the female seizes food brought by the male and gives loud 'dismissal' calls much like the begging calls of fledglings. Some food transfers are aerial.

After the breeding season, some birds (mostly females and juveniles) move into urban areas where they are sometimes seen around aviaries, although they are less prone to conflict with humans than are the two larger goshawks. Sparrowhawks live and breed unobtrusively in some large, well-wooded urban parks.

BREEDING

The laying season is July to December. Pairs nest solitarily. The nest is a platform of sticks 27–32 cm across, 12–15 cm deep, lined with green leaves, and placed 4–39 m above ground in the fork of a living tree. The clutch size is usually three or four eggs, ranging from two to five. Incubation takes 35 days, and the nestling period is about 28–33 days. Success was 1.6 young fledged per clutch laid, and 1.7 per successful nest; nestlings are occasionally taken by goannas. The period of dependence after fledging lasts up to six weeks, after which young disperse. Sexual maturity is attained at one year, with birds sometimes breeding in juvenile plumage.

THREATS AND CONSERVATION

The Collared Sparrowhawk is not globally or nationally threatened. It is widespread and generally uncommon, but can be common in forests in the tropics and subtropics; it is also secretive and probably under-recorded. It has undergone local declines in extensively cleared areas, particularly in south-western Australia where there are few exotic passerines. The thickness of its eggshells was reduced by DDT use in Australia; some local reduction in breeding success was likely. It is possibly affected in south-eastern Australia by the population explosion of the Pied Currawong (*Strepera graculina*), a predator and competitor capable of robbing and injuring adults and killing nestlings. Sparrowhawks sometimes collide with windows.

Chapter 9

Booted eagles, genera *Aquila* and *Hieraaetus*

The species in this group are the raptors usually thought of as 'eagles' in the strict sense, although the smallest species do not fit the popular image in terms of size. They are characterised by powerful bills and feet and feathered or 'booted' tarsi. Australia has one breeding representative from each of two of the two main genera, *Aquila* ('eagle') and *Hieraaetus* ('hawk eagle'): the Wedge-tailed Eagle and Little Eagle. A second species of *Aquila* is a non-breeding visitor to Australia. Both genera occur in the Old World; *Aquila* also reaches North America. Closely related to them are the small, short-winged, tropical forest hawk-eagles of the third main genus, *Spizaetus*, which reaches Wallacea but not Australia. Several monotypic genera in South America and Africa complete the group. As a whole, the group can be characterised as predatory, soaring raptors with broad, rounded wings, which are slotted by emarginated primaries; most have head ornamentation varying from noticeable hackles on the nape to slight or elaborate occipital crests. They are regarded as most closely related to the buzzards (genus *Buteo*), some of which also have feathered tarsi. Richard Holdaway's skeletal comparison challenges the traditional view of relationships among the eagles and buzzards.

The largest booted eagles are among the world's largest raptors (other than vultures). Largest in the group is the Martial Eagle (*Polemaetus bellicosus*) of Africa. Next come two members of the genus *Aquila*: the Golden Eagle (*Aquila chrysaetos*) of the Holarctic and Verreaux's or African Black Eagle (*Aquila verreauxii*). Only slightly smaller is our Wedge-tailed Eagle, followed by the Crowned Eagle (*Stephanoaetus coronatus*) from Africa and several other *Aquila* species that are scarcely smaller again. There is no truth to claims that the Wedge-tailed Eagle is the world's largest eagle.

Australia is also inhabited by one of the smallest booted eagles. Most species in the genus *Hieraaetus* approximate our Little Eagle in size, although Bonelli's Eagle (*Hieraaetus fasciatus*) and the African Hawk-Eagle (*Hieraaetus spilogaster*) are somewhat larger, in the size range of small species of *Aquila*.

Eagles of the genus *Aquila* are usually shades of brown, with conspicuous hackles on the nape, although Verreaux's Eagle, which has striking black-and-white adult plumage, is an exception. They have yelping or barking voices. Members of the genus *Hieraaetus* are often brighter and more variegated, with slight crests and more melodious whistling voices.

Wedge-tailed Eagle
Aquila audax

DESCRIPTION

The specific name ('bold') is a misnomer for this normally shy and wary bird.

The Wedge-tailed Eagle is 85–104 cm long (tail almost half), with a wingspan of 186–227 cm. The average male weighs 2953 g, females 3963 g. It is the largest Australian raptor: a huge dark eagle with long wings, long wedge-shaped tail, and baggy feathered legs. It has a lanky build with a large bill, small head, long neck, and prominent 'shoulders' (carpals) when perched.

The **adult** is sooty or brownish black with tawny hackles on the nape, narrow, mottled grey-brown upperwing bar (occupying less than a quarter of the wing width), brown undertail coverts, and pale bases to the flight feathers (visible on the underwings). The cere is creamy white, the eyes brown, and the feet creamy white. The **juvenile** is dark brown with golden to reddish-brown nape, back, and broad upperwing bar (more than half the width of the wing), barred underwings and tail, and pale undertail coverts. The bulging secondaries produce more curved trailing edges to the wings than in older age classes. The cere is cream, the eyes grey to brown, and the feet cream. **Immatures** (second- to fourth-year birds) are similar but show moult or old and new flight and tail feathers, with the upperwing bar becoming narrower. Older birds (fifth to seventh year) are darker, having a reddish-brown nape, back and upperwing bar, which is narrow still (a quarter to half of the wing width). The **chick** has white down (including on the tarsi), which is long and hair-like on the head, and a prominent black bill.

The Wedge-tailed Eagle is a solitary or gregarious eagle of most terrestrial habitats except intensively settled or cultivated areas. Its flight action is rather loose but with deep and powerful wing-beats; it soars and glides with a modified dihedral, long splayed primaries, and the tail 'dished' from behind (edges curved up in a dihedral) (see figures 9.1 and 9.2). Even in gale-force winds, its gliding flight is stable and controlled. It is usually silent, but sometimes utters weak yelps and squeals, often with a rolling quality.

Figure 9.1 Soaring

Figure 9.2 Gliding

Distant Wedge-tailed Eagles in flight can be confused with other raptors that glide on upswept wings; they are usually distinguished by size, tail shape, plumage, and flight behaviour. Adults with some white in the underwings can be mistaken for the Black-breasted Buzzard, which has distinct white panels in the primaries (both surfaces) and a short, square tail. The juvenile White-bellied Sea-Eagle has shorter, broader, more stiffly upswept wings, a short pale tail, and bare tarsi.

DISTRIBUTION

The Wedge-tailed Eagle occurs throughout mainland Australia and Tasmania, and in the Trans-Fly savannahs of southern New Guinea.

FOOD AND HUNTING

The Wedge-tailed Eagle eats mammals, birds, reptiles, and carrion. In the south it prefers rabbits and hares, in the north young kangaroos and wallabies. Rabbits are important in arid Australia, as are road kills on highways. In tropical rainforest it takes possums. Birds taken commonly include crows, cockatoos, and waterfowl and (rarely) up to the size of cranes and bustards. Reptiles taken are commonly dragons and goannas, rarely snakes.

It forages by low, slow quartering, high soaring or still-hunting from a perch. It seizes prey on the ground or sometimes from the tree canopy after a swift, stealthy glide or dive, which can become a short chase; it rarely takes prey in flight. Occasionally it removes mammals, such as possums, from tree hollows. Pairs or groups attack large prey cooperatively. Individuals gather at carcasses of large animals, dominate other scavengers, and occasionally rob other raptors.

BEHAVIOUR

The Wedge-tailed Eagle is usually seen soaring majestically, high in the air, typically above hill crests, mountain peaks or escarpments, although also over flat plains. It can also be seen sitting prominently on a high perch, such as a tall dead tree or rocky prominence, or on a shrub or low tree in the inland. It is often seen at road kills. Territorial birds soar high, for up to 90 minutes or more, in a display of ownership that might culminate in an Undulating Display (see figure 9.3). A stoop at an intruder could cause the latter to roll and parry with its claws. At all times of year, but particularly in the breeding season, pairs spiral up in a Mutual Soaring Display that can culminate in Rolling and Foot-touching when the male dives at the female. The pair then descends to mate on a branch near the nest. Aerial manoeuvres are sometimes mistaken for attempted copulation on the wing. Mated birds perch together, extend their necks and touch bills or sometimes allopreen. Presentation of nesting material by the male sometimes precedes copulation, and in courtship feeding or provisioning of the nesting female, some food passes are aerial. Territorial Eagles sometimes attack hang-gliders intruding too near their nests.

BREEDING

The laying season is April to September, occasionally earlier in the tropical north. Pairs nest solitarily. The nest is a large platform or pile of sticks typically 70–90 cm across, 30–80 cm deep, and lined with green leaves; it can become 1.8 m wide and 3 m deep with repeated use. It is usually built in a live or dead tree with a com-

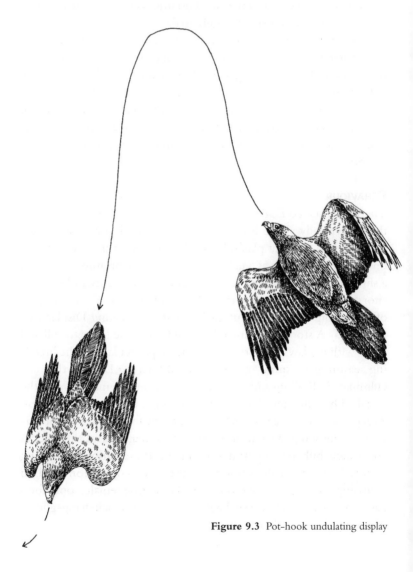

Figure 9.3 Pot-hook undulating display

manding view, 2–73 m above ground: in the tallest tree available, although often near the ground in remote deserts, and typically on a rise or hillside. It occasionally builds nests on cliff ledges, or among rocks and even on the ground on islands in areas inaccessible to humans, and rarely on a structure, such as a power pole or pylon. The clutch size is usually two eggs, ranging from one to three and rarely four; in Tasmania the clutch size is usually one. Incubation takes about 43 days, and the nestling period is 79–95 days. Success varies regionally: in south-western Australia 0.7–1.2 young fledged per clutch laid, 0.19–0.46 young per pair per year; in south-eastern Australia 0.9–1.5 young per clutch laid, 0.6–1.0 young per pair per year; Tasmania 0.8 young per clutch laid, 1.07 per successful nest. The period of dependence after fledging lasts between three and six months, after which young disperse widely (commonly 200 km but up to 800 km). Sexual maturity is reached at three years, and birds will pair in immature plumage, although they seldom breed before acquiring adult plumage at about six or seven years. The oldest banded wild bird was 11 years old; longevity in captivity is up to 40 years.

THREATS AND CONSERVATION

The Wedge-tailed Eagle is not globally or nationally threatened. It is widespread and common on the Australian mainland, despite formerly intense persecution for its supposed impact on domestic livestock (now shown to be negligible). It is still subject to illegal shooting and poisoning. There have been local declines in the south through habitat disturbance, in heavily settled and farmed areas, because the Eagle's intolerance of human activity leads it to abandon its nest. It has benefited elsewhere from thinning of tree cover, introduction of the rabbit, and provision of abundant carrion. The thickness of its eggshells was not significantly reduced by DDT use in Australia. The large, isolated Tasmanian race is classified as Endangered, because its range and population are small (about a hundred breeding pairs) and it has more specific habitat requirements than mainland eagles. Protection of nests by buffer zones and restriction of disturbance in the breeding season are effective conservation measures.

Gurney's Eagle
Aquila gurneyi

DESCRIPTION

This species (named after J.H. Gurney, a nineteenth-century British ornithologist) is endemic to New Guinea and the Moluccas. It has been recorded on Torres Strait islands in Australian territory (close to the New Guinea mainland), where it might be resident, and on the tip of Cape York Peninsula, where it is presumably a vagrant. It is not known to breed in Australian territory. Its nesting habits have not been described.

Gurney's Eagle is 74–86 cm long (tail almost half), with a wingspan of about 170–190 cm. A female weighed 3060 g. It is similar in size to the Wedge-tailed Eagle and White-bellied Sea-Eagle. It is a large, dark eagle with long, broad wings, long rounded tail, and slim feathered legs.

The **adult** is entirely blackish brown with a paler upperwing band and mottled grey on the undersides of the remiges and tail feathers. The cere is creamy grey, the eyes brown to dark yellow, and the feet creamy to dull yellow. The **juvenile** is browner than the adult: variegated grey, light brown, and dark brown on the upperparts, with a dull rufous head and pale grey upperwing band; rufous, dark-streaked underparts; and paler tarsal feathering and vent. The cere is bluish white, the eyes brown to hazel, and the feet white. The downy **chick** has not been described, although presumably it is white with down-clad tarsi as other *Aquila*.

Gurney's Eagle is a solitary eagle of tropical forests—mainly lowland and coastal rainforest. Its flight action is typical of the genus: powerful and heavy. It glides on broad, flat to slightly raised, almost straight and parallel-sided wings with bulging secondaries, fingered primaries, and a long, ample tail with a rounded to slightly wedge-shaped tip (see figure 9.4). Its voice is a slightly nasal, downslurred piping.

Gurney's Eagle can be confused with the Wedge-tailed Eagle but has yellow feet and a rounded tail, and soars on almost flat (not obviously upswept) wings. The juvenile White-bellied Sea-Eagle has shorter, broader wings with curved trailing edges, a much shorter, pale tail, bare tarsi, and stiffly upswept wings in gliding or soaring flight.

Figure 9.4a Soaring and gliding

Figure 9.4b Gliding

Little Eagle
Hieraaetus morphnoides

DESCRIPTION

The specific name ('eagle-like'), in its diminutive form, implies a likeness to larger eagles.

The Little Eagle is 45–55 cm long (tail less than half), with a wingspan of 110–136 cm. The average weight of males is 635 g, females 1046 g. It is similar in size to the Whistling Kite but more robust, with shorter wings and tail, a short, broad head, and a stubby bill. It is a powerful, stocky small eagle with a short occipital crest, feathered legs, and dark upperparts with a narrow, pale upperwing bar. The plumage is polymorphic: pale birds have almost white underparts with a striking underwing pattern; dark birds are brown with pale inner primaries. A rare rufous variant of the pale morph has much redder underparts and heavier streaking below.

Polymorphism occurs in both sexes and can occur in a brood of juveniles. The **adult pale morph** has dark brown upperparts with a conspicuous pale upperwing band and scapulars. The head and underparts are sandy or pale rufous, white on the belly, with heavy black streaks on the crown and crest and dark shaft streaks on the breast. Dark facial streaks give some birds a hooded appearance. The underwing pattern is distinctive: a rufous leading edge, pale oblique band, grey-barred secondaries, pale inner primaries, and barred, black-tipped outer primaries. The tail is barred. The cere is pale grey or cream, the eyes reddish brown to orange-brown, and the feet pale grey or cream. Females tend to have more rufous smudging and heavier dark streaks on the breast than do males. A rare rufous variant has the head and entire underparts richer rufous, with heavier black streaks. The **juvenile pale morph** is similar to the adult pale morph but richer rufous on the head and underparts with little or no dark streaking, a darker, less contrasting upperwing band and (often) rufous spots on the upperwings. It has a thin pale line along the centre of the spread upperwing, formed by pale tips to the greater coverts, and thin translucent trailing edges to wings and tail. The cere is cream to pale yellow, the eyes brown, and the feet cream to pale yellow. The **adult dark morph** resembles the adult pale morph above. The head and underparts are light to dark brown with heavy black streaks. The underwings are wholly dark except for pale inner primaries. **Females** tend to be darker and more heavily streaked than males, but occasional exceptionally dark birds of both sexes occur. The **juvenile dark morph** is similar to the juvenile pale morph but is darker rufous or reddish brown on the head and underparts, sometimes with black (although narrow) streaks on the breast; the underwing pattern is as for the adult dark morph. The **chick** has white down (including on the tarsi), which is long and hair-like on the head and buff on the belly; it also has a dark 'mask' around the eyes.

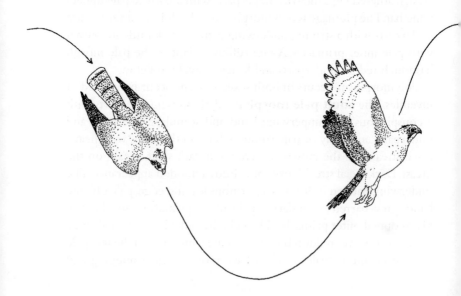

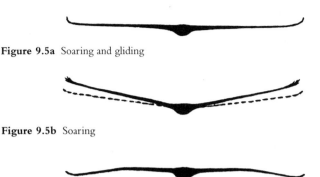

Figure 9.5a Soaring and gliding

Figure 9.5b Soaring

Figure 9.5c Gliding

The Little Eagle is a solitary raptor of most wooded habitats, although it usually avoids dense forest. It is characteristic of woodland in rough hilly country or of river gums in the inland. Its flight action is powerful and fluid, somewhat laboured, although the bird is capable of swift dives with closed wings. In soaring and gliding flight it is stable and controlled, even in strong winds, showing the pale upperwing band and a rather short, square tail. It glides on flat wings with the primaries slightly lowered (see figure 9.5). It soars

Figure 9.6 Undulating display

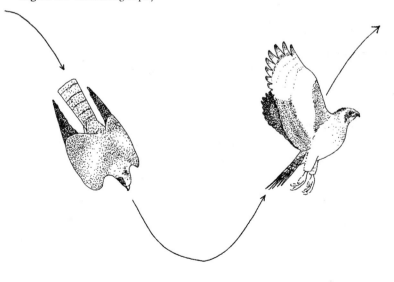

with the wings held straight out from the body, the leading and trailing edges parallel, and the primaries sometimes slightly swept back. The wings might be slightly raised when soaring to gain height, and the tail is sometimes widely fanned. It has a characteristic wind-hanging hunting attitude: splayed wings held slightly forward, alulae projecting, and tail fanned. It can drop towards the ground in several stages in this posture. Its most common call is distinctive: a loud, rapid, excited two- or three-note whistle, often heard before the high-soaring bird is seen. Its display flight is also distinctive: a series of aerial undulations with or without wing-beats on the upswing, producing an obvious flash pattern in pale birds, with the whistling call given loudly and frequently (see figure 9.6).

The Little Eagle can be confused with several species. The Whistling Kite has a smaller head, slender body; longer, more splayed primaries, curved trailing edges to the wings, longer rounded tail, bare tarsi, and small feet. The juvenile Brahminy Kite lacks the upperwing band and has an indistinct underwing pattern, curved trailing edges to the wings, a rounded tail, and bare tarsi. The Square-tailed Kite is slender and has a much more attenuated appearance with longer, barred, more splayed primaries, upswept wings, longer tail, and small feet. The Black Kite is more slender with longer primaries, longer forked tail, and small feet. The Little Eagle is sometimes mistaken for the Black-breasted Buzzard or Red Goshawk (for differences, see the accounts for those species).

DISTRIBUTION
The Little Eagle occurs throughout mainland Australia and in New Guinea.

FOOD AND HUNTING
The Little Eagle eats mammals, birds, reptiles, occasionally large insects and carrion, and rarely fish (the last possibly robbed from the Whistling Kite). In the south it prefers young rabbits, in the north birds, and in the arid zone lizards. It catches mammals up to 1.5 kg in weight; birds commonly taken are parrots and passerines, rarely up to 1 kg (ducks, crows); reptiles taken are commonly dragons, small goannas, and large skinks, but rarely snakes. It forages by quartering

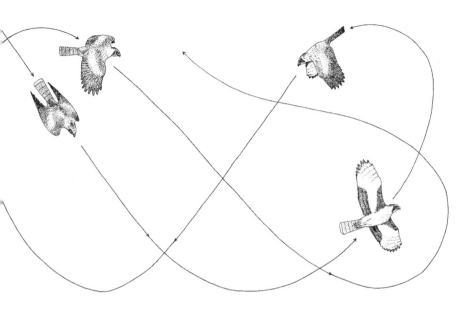

Figure 9.7 Undulating display

and high soaring, by low flights between perches, or by still-hunting from a perch. It seizes prey on the ground by a glide or dive, sometimes by a rapid stoop to the tree canopy. It rarely takes prey in flight.

BEHAVIOUR

The Little Eagle spends much time soaring high, often near or above the limit of human vision, when its calls can be heard before the bird is seen. Its small size, and habit of perching quietly in a living or dead tree, mean that it is easily overlooked. At any time of year, but particularly during the breeding season, territorial birds soar in a special posture with the tail furled and primaries turned back, and perform a vigorous Undulating Display, all the while with loud calling (see figure 9.7). Members of a pair soar together, the male's approach or dive eliciting Talon-presentation by the female (see figure 9.8), or some tumbling by both birds. When soaring to gain

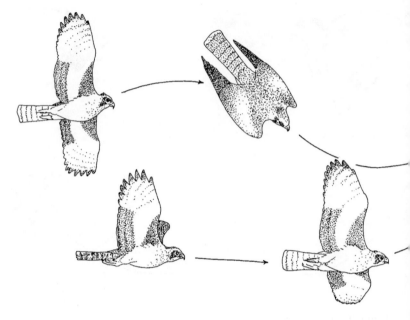

Figure 9.8 Mutual soaring and talon presentation

height, wing-beats can be short and goshawk-like. Solitary or Mutual Soaring sometimes culminates in a long dive to the tree canopy or nesting site. Food-begging females squeal and chitter loudly at the male.

BREEDING

The laying season varies with latitude: longer in the tropical far north, occupying the dry season (March to September), shorter in the centre and south (usually August to October; rarely starting in May or extending to December for replacement clutches).

Pairs nest solitarily. The nest is a platform of sticks 60–75 cm across, about 30 cm deep, lined with green leaves, and placed 5–45 m above ground in the fork of a living tree. The clutch size is usually two eggs, but it ranges from one to three. Incubation takes about 36 days, and the nestling period is 54–66 days. Success varies regionally and according to weather (lower in wet or drought years):

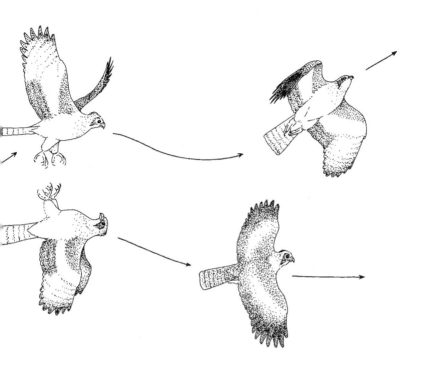

0.5–1.0 young fledged per pair per year; 0.8–0.9 young per clutch laid; 1.1 young per successful nest. The period of dependence after fledging lasts about two months, after which young disperse or migrate up to 3000 km. The oldest banded wild bird was 26 years old.

THREATS AND CONSERVATION

The Little Eagle is not globally or nationally threatened. It is common and widespread in Australia, where presumably it has benefited from thinning of tree cover and introduction of the rabbit. It has increased in range and numbers in coastal south-eastern Australia. The thickness of its eggshells was not significantly reduced by DDT use in Australia. It is not easily disturbed at the nest and is seldom persecuted. It is possibly affected, but only slightly, by extensive habitat clearance or the decline of rural trees ('eucalypt dieback') in agricultural areas.

Chapter 10

Falcons, genus *Falco*

The six Australian members of the Falconidae all belong to the cosmopolitan genus *Falco* ('falcon'). Penny Olsen has proposed a radical new concept of relationships within the genus and a Gondwanan origin for some Australasian species. One, the Nankeen Kestrel, is a member of the widespread group of kestrels or small hovering falcons that apparently originated in Africa. The Nankeen, and its close relative, the Moluccan Kestrel (*Falco moluccensis*), stand slightly apart from the other kestrels; their ancestor probably colonised Australasia via South-East Asia. The Peregrine Falcon, having differentiated little from other forms within the species elsewhere in the world, is an even more recent colonist. The other four falcons represent an old endemic radiation dating from Gondwanan times. The Grey and Black Falcons are closely related to another pair of species, the Australian Hobby and the Oriental Hobby (*Falco severus*); all four constitute the Australasian hobbies. They are most closely related to the Gondwanan hobbies: a group consisting of the Brown Falcon and its close relative, the New Zealand Falcon (*Falco novaeseelandiae*), which are in turn most closely related to a trio of species, the Sooty Falcon (*Falco concolor*), Eleonora's Falcon (*Falco eleonorae*) and Bat Falcon (*Falco rufigularis*). The Sooty and Eleonora's Falcons are African (in the zoogeographical sense), and the Bat Falcon is South American. These suggested relationships add weight to the argument for a Gondwanan origin for the Falconiformes and some lineages within the Accipitridae and Falconidae.

The falcons are characterised by their dark facial or head markings, dark eyes, conspicuous and often brightly coloured orbital skin and other bare parts, long pointed wings, and usually rapid flight. At close range, the tomial 'tooth' on each side of the cutting edge of the upper mandible, near the tip, is evident.

Brown Falcon
Falco berigora

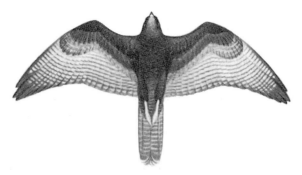

DESCRIPTION

The specific name is an Aboriginal name for this falcon. The species is characterised by long legs and short toes with scutellate scaling on the front of the tarsi and reticulate scaling behind; the scales are large, rough and overlapping.

The Brown Falcon is 41–51 cm long (tail almost half), with a wingspan of 89–109 cm. The average weight of males is 476 g, females 610 g. It is larger than the Nankeen Kestrel and similar in size to the Brown Goshawk but with longer wings. It is a medium-sized, rather scruffy, loose-plumaged falcon with a large head, rounded shoulders, long legs, and a raucous voice. It tends to be sluggish, perching conspicuously with a pot-bellied profile. Its flapping flight is slow and heavy, with glides on raised wings, at times like the flight of a harrier. It also hovers, rather clumsily compared with the Black-shouldered Kite or Nankeen Kestrel. The wing tips are blunt and flexible, and the tail is rounded. It is polymorphic and its plumage varies greatly, from uniformly dark like the Black Falcon to pale like the Nankeen Kestrel; dark birds have barred underwings and tail.

Polymorphism occurs in both sexes. The ratio of colour morphs varies geographically, with mixing of morphs: only the brown morph occurs in Tasmania and the brown morph predominates in south-eastern Australia; the rufous morph is common throughout the inland and Western Australia; the dark morph is common in northern Australia. The **adult brown morph** has a pale, or sometimes grey, forehead and a pale face with dark malar stripe and ear patch. The upperparts are brown with rufous spots and bars. The underparts are white with fine dark shaft streaks, brown thighs, and brown spots on the flanks; the undertail coverts are white. The underwings and tail are barred. Some birds have brown or mottled brown breast and flanks. The cere and eye-ring are pale grey to pale blue-grey or (rarely) yellow, the eyes brown or (rarely) hazel, and the legs and feet pale blue-grey to pale grey or (rarely) dull yellow. The **adult rufous morph** has upperparts similar to the brown morph but is more reddish brown; the underparts are rufous with fine dark shaft streaks, a variable white centre to the belly, white spots on the flanks, and barred undertail coverts. The **adult dark morph** is uniformly dark brown with an indistinct malar stripe, indistinct russet spots on the flanks, russet barring on the tail, and barred or dark undertail coverts. The underside of the remiges and undertail are pale and strongly barred. The **juveniles** of all morphs are rather uniformly dark. The cere and eye-ring are pale grey or blue-grey, the eyes brown, and the legs and feet pale blue-grey to pale grey. The **brown morph** is dark brown with a buff face, double cheek-mark, broad buff collar, fine rufous edging to the feathers of the back and wings, buff or cream undertail coverts, and mottling down the centre of the belly; it has no spots on the flanks. The underwings and tail are barred. In many birds the underparts become whiter with age, starting with the undertail coverts and progressing forward, more rapidly and completely in males. Some birds retain brown underparts to adulthood. The **juvenile rufous morph** is similar to the brown morph but more russet; it lacks spots on the flanks and a pale collar. The **juvenile dark morph** is similar to the dark adult, but without spots on the flanks; it lacks a pale collar. The **chick** has pale rufous first down and pale grey second down.

The Brown Falcon is solitary or loosely gregarious. It is an excitable, highly vocal falcon of most open habitats but avoids dense forest. Its flight action is loose and rowing, with heavy, rather slow wing-beats, and its flight is often erratic with jinking and side-slipping. It can fly swiftly in pursuit with short, stiff wing-beats, not flickering like the faster falcons and with the outer-wings swept back. It glides on raised wings and soars with rounded wing tips upswept, showing curved trailing edges to the wings and a rounded tail, which is often fanned (see figures 10.1 and 10.2). It hovers or hangs on the wind. At rest it sits high on long legs, with the wing tips reaching to or just beyond the tail tip. It often gives cackling, chattering, and screeching calls.

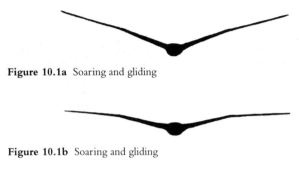

Figure 10.1a Soaring and gliding

Figure 10.1b Soaring and gliding

Figure 10.1c Soaring and gliding

Pale adult Brown Falcons can be confused with the Nankeen Kestrel, but are larger and browner with long legs and brown thighs; they hover with slow wing-beats and glide on raised wings. The Red Goshawk can be mistaken for the rufous morph but is more robust, with a crested nape, greyer flight and tail feathers, more massive legs and feet, squarer tail, and faster flight; it lacks the dark facial marks. The dark Brown Falcon is often mistaken for the Black Falcon, which differs by more pointed wings, faster flight, square-tipped tail, and short legs, and gliding on drooped wings (for further differences, see the account for that species) (see figures 10.2 and 10.3). Large Brown Falcons, particularly juveniles with rather broad wings and buoyant flight, can be confused with the Swamp Harrier, but Falcons have a double cheek mark and more tapered, less upswept wings, and lack a pale rump.

Figures 10.2 Soaring (a) Brown Falcon (b) Black Falcon

Figures 10.3 Gliding (a) Brown Falcon (b) Black Falcon

DISTRIBUTION

The Brown Falcon occurs throughout mainland Australia and Tasmania, and in New Guinea.

FOOD AND HUNTING

The Brown Falcon eats mammals, birds, reptiles (including snakes), amphibians, arthropods, carrion and (rarely) fish. It forages mostly by still-hunting from an exposed perch; also by quartering and hovering, low fast flight, or by soaring. It seizes prey on the ground by a glide, dive or direct flying attack that might become a short chase (see figure 10.4). It pursues insects on foot and robs other raptors. It also follows fires, livestock and other animals, and farm machinery for flushed prey. Members of a pair sometimes hunt cooperatively.

BEHAVIOUR

The Brown Falcon is a common roadside raptor, typically seen perched on fence posts, utility poles, shrubs, or the top of dead trees. It also perches on wires, balancing precariously while fanning its

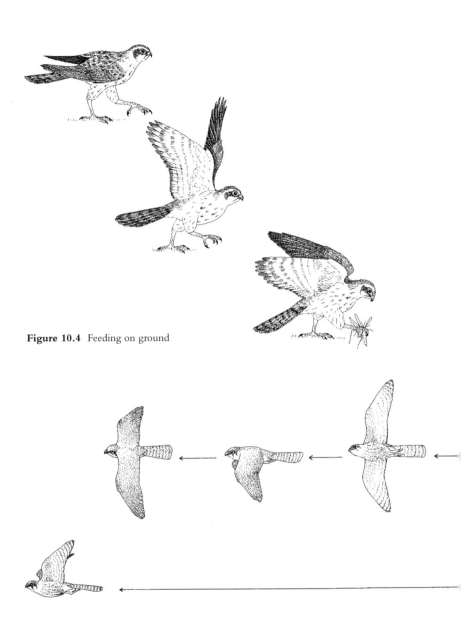

Figure 10.4 Feeding on ground

Figure 10.5 Fast chasing flights in courtship

tail. In flight it often gives loud, raucous, and crowing calls. At the start of the breeding season, males soar and perform erratic or zigzagging diving displays with flickering flight and cackling calls. Members of a pair soar together with laboured wing-beats and much squawking and cackling. The male might dive at the female, who will roll to present her talons, or they might perform a fast, low-level chase with erratic, flickering flight that culminates in perching and greeting with a Bowing Display, chattering calls, and perhaps allopreening (see figure 10.5). Food presentation by the male takes place either at a perch or aerially.

BREEDING

The laying season is April to September in northern Australia and August to October in the centre and south. Pairs breed solitarily. The Brown Falcon uses the old stick nest of another raptor or corvid, usually situated in a tree, rarely on a tree-fern, vine, artificial structure, cliff or termitarium, 4–30 m above the ground. The clutch size is usually two or three eggs, ranging from one to five. Incubation takes about 33 days, and the nestling period is 36–42 days. Success has been measured variously as 69% fledging success (of eggs laid), 1.77 young fledged per nest and 2.39 per successful nest; 2.2 young raised per clutch laid, 2.3–2.4 young per successful nest and 1.7 young per territorial pair. The period of dependence after fledging lasts up to six weeks, after which young disperse widely (up to 2000 km). Some Tasmanian juveniles migrate to the mainland, and some southern mainland birds migrate to the tropics. Age at first breeding is 3 years for males and 2 years for females. The oldest banded bird was 11 years; longevity in captivity is up to 20 years.

THREATS AND CONSERVATION

The Brown Falcon is not globally or nationally threatened. It is common and widespread; it benefits from most agricultural activities, but is sometimes shot or trapped. The thickness of its eggshells was not significantly reduced by DDT use in Australia.

Australian Hobby (Little Falcon)
Falco longipennis

DESCRIPTION

The specific name ('long feather') alludes to the long wings, specifically the primary feathers. The species is characterised by long toes and a reticulate pattern of tarsal scalation.

The Australian Hobby is 30–35 cm long (tail almost half), with a wingspan of 66–87 cm. The average male weighs 217 g, females 270 g. It is similar in size to the Nankeen Kestrel but more robust, with more pointed, rakish wings and a slightly shorter, square tail. It is a small, dark, long-winged falcon with a slim build, dashing flight, and elegant shape, reminiscent of a swift.

The **adult** has a black crown and cheeks, which form a partial helmet, and cream or buff forehead, throat and half-collar. The upperparts are slaty blue-grey to dark slate-grey, with black wing tips. The underparts are buff to rufous with fine dark streaks on the breast and dark spots on the flanks. The underwings and tail are finely barred. Birds from southern humid regions are darker than inland and northern birds. The cere is pale yellowish grey to pale yellow, the eye-ring pale blue, the eyes brown, and the feet dull yellow. The **juvenile** is darker and browner than the adult, with a richer rufous suffusion to the head and underparts, rufous feather edges on the back and wings, and bars on the tail; it lacks spots on the flanks. The cere and eye-ring are pale blue, the eyes brown, and the feet pale yellow. The **chick** has white to cream or sometimes pale rusty down.

The Australian Hobby is a solitary and aggressive falcon of most open habitats, including vegetated urban areas, but is rarely recorded around cliffs and escarpments; it is characteristic of open woodland and watercourses. Its flight action is dashing and flickering when hunting but otherwise more leisurely. It glides on flat or slightly drooped wings, with the carpals flexed and outerwings swept back. It soars on flat outstretched wings somewhat turned back at the carpals, with curved trailing edges (see figure 10.6). Its flight is often low and fast, zigzagging over or between trees or low over open ground. It is acrobatic in pursuit of small birds, bats and flying insects. It attacks fleeing birds in a series of short, shallow stoops with towering, flickering flight on the upswing. It often harasses large birds but is incapable of killing them. It seldom lands on the ground, instead preferring prey that it can eat on the wing or on a high perch. At rest it shows a slim build, small bill and feet, and long wings reaching to or just beyond the tail tip. Its most commonly heard call is a sharp, peevish chatter like that of the Nankeen Kestrel but reedier and hoarser; also squeaky chittering and ticking, and a loud chuckling in winnowing display flight.

Figure 10.6 Soaring and gliding

The Australian Hobby is most likely to be confused with the Peregrine Falcon, which is a more heavily built and powerful flier with a large head and bill, full helmet, deep chest, relatively shorter and broader wings and tail, and large feet (see figures 10.11 and 10.12). Pale inland Hobbies can be confused with the Grey Falcon, which lacks dark head markings and pale rufous underparts, and has bright orange-yellow bare parts. The Nankeen Kestrel occasionally appears similarly swift-like but is paler, with a longer rounded tail and less sharply pointed wings. The Collared Sparrowhawk has shorter, broader, rounded wings (not reaching the tail tip at rest) and long legs.

DISTRIBUTION

The Australian Hobby occurs throughout Australia, although sparsely in Tasmania. Migrating individuals occur in New Guinea and occasionally the Moluccas. An endemic, resident race (*hanieli*) occurs in the Lesser Sunda Islands.

FOOD AND HUNTING

The Australian Hobby eats small birds, insectivorous bats, and flying insects. It captures mostly terrestrial flocking birds, such as larks, pipits, grassfinches, doves, and small parrots but also takes aerial species such as swallows. In farmland and towns, its prey is mainly introduced sparrows and starlings. Most prey weighs less than 100 g and is rarely more than 200 g. Its hunting is diurnal, crepuscular, and sometimes nocturnal by artificial light. It forages by low fast flight, still-hunting from a prominent perch, or high quartering. It seizes prey in flight by a shallow stoop or direct flying attack that becomes a vigorous chase. It hawks flying insects and occasionally steals mice from the Nankeen Kestrel.

BEHAVIOUR

The Australian Hobby is typically seen dashing past at low level or perched on a prominence, such as a high dead branch, television aerial, mast or tower. It is active well into dusk or even after dark. Members of a pair engage in noisy display flights, with soaring, chasing, and mock attack-and-parry accompanied by chittering calls. Food passes from male to female are often aerial. Females also perform a Bowing Display on the nest, with loud chattering calls. Some birds breeding at high latitudes and altitudes are migratory, leaving in autumn; there is a corresponding winter influx to the tropics.

BREEDING

The laying season is August to January. Pairs nest solitarily. The Hobby uses an old or usurped stick nest of another species, typically a corvid, high (more than 10 m) above the ground, in the top of a tree or electricity pylon. The clutch size is usually two or three eggs, ranging from two to four. Incubation takes about 35 days, and the nestling period is 34–38 days. Success was 2.75 young per year for a pair over 4 years; successful broods commonly comprise 2–3 young. The period of dependence after fledging lasts up to three months, after which young disperse or migrate widely (up to 900 km has been recorded). The oldest recovered banded bird was 7 years old.

THREATS AND CONSERVATION

The Australian Hobby is not globally or nationally threatened. It is fairly common and widespread, including in cities and towns. Its population is probably stable, and it has benefited from the introduction of prey species in some parts of Australia. The thickness of its eggshells was reduced by DDT use here, and some local declines in breeding success probably occurred in southern agricultural areas. It is seldom shot.

Grey Falcon
Falco hypoleucos

DESCRIPTION

The specific name ('under white' or 'less than white') refers to the white or pale underparts. The species is characterised by finely reticulate tarsal scalation and long toes.

The Grey Falcon is 33–43 cm long (tail less than half), with a wingspan of 86–97 cm. A male weighed 336 g, and females average 567 g. It is a medium-sized, pale grey falcon with long pointed wings and short tail. In flight, the black wing tips contrast with the grey upperparts and pale underparts. When perched, it appears broad-shouldered and short-legged, with striking orange-yellow cere, eye-ring, and feet.

The **adult** has grey upperparts with a faint dark malar stripe and black wing tips. The underparts are white or very pale grey with faint shaft streaks. The underwings and tail are finely barred. The cere and eye-ring are orange-yellow, the eyes brown, and the feet orange-yellow. The **juvenile**'s upperparts are somewhat darker than those of the adult, with a more distinct malar stripe and pale brown edges to the feathers of the mantle. The underparts are white with fine dark streaks on the breast and dark spots on the flanks. The cere and eye-ring are pale blue-grey, the eyes brown, and the feet pale yellow. The **chick** has white down.

The Grey Falcon is a solitary desert falcon, sometimes seen in pairs or family groups. It is characteristic of shrubland, grassland, and wooded watercourses of the arid zone, but visits northern coasts and occasionally appears in open areas in humid southern regions. Its flight action is a kestrel-like winnowing, often leisurely with shallow, fluid wing-beats, although sometimes with higher beats like a fast version of the Brown Falcon. It glides on flat wings and soars with the wing tips slightly upswept (see figure 10.7). In soaring flight the wings are held somewhat stiffly forward, with slightly rounded tips. At rest it sits low on short legs, showing long wings that reach the tail tip. Its calls consist of hoarse chattering, clucking, and whining.

Figure 10.7a Soaring and gliding

Figure 10.7b Soaring and gliding

Several other pale or grey raptors can be confused with the Grey Falcon. The *Elanus* kites have black shoulders and underwing markings and white tails, and consistently hover and glide on upswept wings. The Grey Goshawk has short, broad, rounded wings, long legs, and heavier flight, and glides on bowed wings. Pale-backed adult Brown Goshawks have blunt wings, long tail, rufous underparts, long legs, and heavier flight. Sooty-grey individuals of the Black Falcon are darker and more uniformly coloured with a longer tail and drooped wings in glide. They lack the pale underparts and yellow bare parts. The adult Peregrine Falcon has a full black helmet. The pale inland Australian Hobby is slimmer with a partial helmet (dark cheeks) and pale rufous underparts, and lacks the bright orange-yellow bare parts. Very pale Brown Falcons have browner upperparts, long legs, and slower, heavier flight. They lack the bright orange-yellow soft parts. The Nankeen Kestrel is pale from below and males have a grey head, but it has narrower wings and a longer tail with a black subterminal band. It often hovers.

DISTRIBUTION

The Grey Falcon occurs sparsely in the interior and north of the Australian mainland. Two records exist of vagrants in the savannahs of southern New Guinea.

FOOD AND HUNTING

The Grey Falcon eats mostly pigeons and parrots but also other birds, small mammals, lizards, and large insects. It forages by low fast flight, quartering, and high soaring, or still-hunting from a perch. It seizes or strikes prey in flight by a stoop or direct flying attack, or glides from a perch to take prey on the ground.

BEHAVIOUR

The Grey Falcon is a rarely seen bird of inland Australia. Even in its normal haunts it can be quiet, unobtrusive, and difficult to detect. It might perch cryptically on a dead tree or among the foliage of a living tree. It can also be seen in flight, which is swift and hobby-like, at treetop level, making sudden changes of direction and covering a wide area in search or pursuit of birds. It sometimes sweeps

Grey Goshawk *Accipiter novaehollandiae*, nominate *novaehollandiae* **1**, **2**, **3** Adult males, grey morph; **4**, **5** Adult females, white morph; **6**, **7**, **8** Juvenile females, grey morph

A Little Eagle in flight *(Nicholas Birks)*
Red Goshawk *(David Baker-Gabb)*

A two-year-old female Wedge-tailed Eagle about to land *(Nicholas Birks)*

A young male Wedge-tailed Eagle approaching a dead kangaroo *(Nicholas Birks)*

Wedge-tailed Eagle *Aquila audax* **1** Adult male, nominate *audax*; **2** Adult female, nominate *audax*, pale; **3** Juvenile, nominate *audax*; **4** Juvenile, subspecies *fleayi*; Gurney's Eagle *Aquila gurneyi* **5** Adult; **6** Juvenile

An adult male Grey Falcon feeding *(Nicholas Birks)*
A female Brown Falcon in hovering flight *(Nicholas Birks)*

Wedge-tailed Eagle *Aquila audax* **1**, **2** Adult males, nominate *audax*; **3**, **4** Adult females, nominate *audax*; **5**, **6** Juveniles, nominate *audax*, fresh plumage; **7** Juvenile, worn; Gurney's Eagle *Aquila gurneyi* **8**, **9** Adults; **10**, **11** Juveniles

Black Falcon *Falco subniger* **1** Adult male; **2** Adult female, light bird; **3** Juvenile male; Brown Falcon *Falco berigora* **4** Adult male, dark morph; **5** Adult female, brown morph (dark bird); **6** Adult male, brown morph (pale bird); **7** Adult male, rufous morph; **8** Adult male, rufous morph (pale bird); **9** Juvenile male, brown morph (pale bird); **10** Juvenile male, dark morph

Black Falcon *Falco subniger* **1** Adult; **2** Adult; **3** Adult, light bird; Brown Falcon *Falco berigora*
4 Adult, dark morph; **5** Adult, dark morph; **6** Adult, brown morph; **7** Adult, brown
morph; **8** Adult, rufous morph; **9** Adult, rufous morph; **10** Juvenile, rufous morph;
11 Juvenile, brown morph; **12** Juvenile, brown morph

A juvenile male Australian Hobby *(Nicholas Birks)*

A juvenile Australian Hobby with prey *(Nicholas Birks)*

Peregrine Falcon *Falco peregrinus* (subspecies *macropus*) **1** Adult male, rufous; **2** Adult female;
3 Juvenile; Grey Falcon *Falco hypoleucos* **4** Adult female; **5** Juvenile; Australian Hobby *Falco
longipennis* **6** Adult male, dark; **7** Adult female, pale; **8** Juvenile male, dark; Nankeen Kestrel
Falco cenchroides **9** Adult male; **10** Adult female; **11** Juvenile

Grey Falcon *Falco hypoleucos* **1** Adult; **2** Juvenile; Peregrine Falcon *Falco peregrinus* (subspecies *macropus*) **3** Adult; **4** Juvenile; Australian Hobby *Falco longipennis* **5** Adult, dark; **6** Juvenile, pale; Nankeen Kestrel *Falco cenchroides* (nominate *cenchroides*) **7** Adult male; **8** Adult female; **9** Juvenile

A female Peregrine Falcon in flight *(Nicholas Birks)*

A sub-adult Nankeen Kestrel in hovering flight *(Nicholas Birks)*

rapidly over waterholes at low level. In display, solitary birds or pairs soar and call, sometimes flying with exaggerated wing-beats or wings raised above the body, like a Brown Falcon. Male and female engage in mock attack-and-parry manoeuvres. Greeting at the nest is accompanied by clucking calls; food passes take place aerially and on the nest or a perch. Pairs are usually resident except in drought when they disperse to the coast or inland refugia. Some birds, possibly mainly juveniles, migrate to winter in northern Australia.

BREEDING

The laying season is June to November. Pairs nest solitarily. The Grey Falcon uses the large stick nest of another bird, typically a corvid, in the top of an emergent living tree 9–25 m above ground. The clutch size is usually two or three eggs but ranges from two to four. Incubation takes about 35 days, and the nestling period is about 41–52 days. The period of dependence after fledging is not known, but is probably long (several months). Family groups persist almost until the next breeding season.

THREATS AND CONSERVATION

The Grey Falcon is globally threatened and classed as Rare (recently revised to Vulnerable). It is scarce and possibly declining; its breeding range has contracted somewhat to the arid zone. Its population has been estimated as about a thousand breeding pairs. It is affected by habitat degradation, and the Peregrine Falcon, which is benefiting from the increase in numbers of Galahs and feral pigeons, is possibly replacing it in the agricultural semi-arid zone. The eggshell thickness of Grey Falcons was reduced by DDT use in Australia; local declines in breeding success in the south are likely to have occurred. Eggs are collected illegally, and nestlings might also be taken for clandestine and illegal falconry activities. Conservation measures required include a population survey and research into the Falcon's biology and ecology.

Black Falcon
Falco subniger

DESCRIPTION

The specific name ('somewhat black') alludes to the bird's dark plumage, which is dark brown rather than truly black. The species is characterised by short tarsi with finely reticulate scalation and moderately long toes.

The Black Falcon is 45–56 cm long (tail about half), with a wingspan of 97–115 cm. Males weigh 638 g on average, females 835 g. It is the largest Australian falcon, similar in body size to the Peregrine Falcon but with longer, broader wings, and a longer tail. It is a sleek, fierce, uniformly dark falcon with a small head, square shoulders, pointed wings, and short legs. It is highly aerial, often soars, and is fast and agile in a stoop or pursuit.

The **adult** is dark brown to sooty black, darkest when the plumage has been recently replaced, with two-toned underwings (remiges slightly paler) and an inconspicuous dark malar stripe. Variable pale highlights to the plumage are sometimes present: white chin, buff forehead and cheeks, white speckling on the breast, spots on the underwing coverts and bars on the undertail coverts, and faint narrow bars on the underwings and undertail. The cere and eye-ring are pale grey, the eyes brown, and the feet pale grey. The **juvenile** is usually darker than the adult, most noticeably so against its own faded parents; it has narrow pale edges to the feathers of the back and wings. Pale markings (white chin, buff streaks on the fore-head, faint barring on flight and tail feathers) are present in at least some birds. The cere and eye-ring are pale blue-grey, the eyes brown, and the feet pale blue-grey. A rare colour form of adults and juve-niles is sooty grey on the upperparts. The **chick** has white down.

The Black Falcon is a solitary, active, aggressive falcon character-istic of open plains and sparse woodland and shrubland of the Aus-tralian interior, sometimes appearing in open areas near the coast. Its flight action is a direct kestrel-like winnowing with short, stiff wing-beats, or slower, more fluid and crow-like. Pursuit flight is powerful with rapid, vigorous wing-beats. It soars and glides on sharply pointed, slightly drooped wings with the carpals held for-ward and straight trailing edges (see figures 10.8 and 10.9); the long square tail is usually held furled but sometimes fanned to reveal stepped outer edges (where the outermost feather is shorter than the next). The pale chin is often obvious. It soars for long periods without flapping, and stoops to level out and chase birds or snatch prey from the ground without landing. At rest it sits low on short legs, with the long tail extending just beyond the wing tips. It is usu-ally silent, but its calls include a deep, harsh chatter, slow whining, and a soft whistle.

The Black Falcon is most likely to be confused with the dark Brown Falcon. They are readily distinguished in flight by their flap-ping styles (rapid and shallow in Black; heavy, rowing, and often erratic in Brown), wing attitude when soaring or gliding (drooped

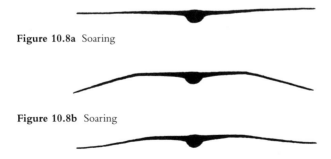

Figure 10.8a Soaring

Figure 10.8b Soaring

Figure 10.9 Gliding

in Black, raised in Brown) and tail shape (square and furled in Black, rounded and often fanned in Brown) (see figures 10.2 and 10.3). When perched, they are distinguished by shape and posture: Black has a small head, broad square shoulders, and short legs with large feet; Brown has a large head, rounded shoulders, pot-bellied profile, and long legs with stubby toes. Sooty-grey Black Falcons can be confused with the Grey Falcon, which is paler (conspicuously so on the underparts) with a shorter tail, flat or slightly upswept wings in flight, and orange-yellow cere, eye-ring and feet. The Peregrine Falcon has a shorter tail and flat wings in glide. Single Black Falcons can be overlooked in flocks of Black Kites, which have loose plumage, separated primaries, pale upperwing bands, and forked tails, which are often fanned. The Black Falcon can also be confused with crows or ravens, which have projecting bills and broader, more rounded wings. Distant soaring male Darters (*Anhinga melanogaster*) resemble the Black Falcon unless the long neck and bill are seen.

DISTRIBUTION

The Black Falcon occurs over much of the Australian mainland except densely forested parts, although it is sparse in the south-western third and coastal south-east.

FOOD AND HUNTING

The Black Falcon eats mammals, birds, large insects, and carrion; rarely reptiles. It commonly takes young rabbits and rats, and birds such as parrots, button-quails, quails, larks, and pipits; rarely it takes larger birds such as herons and waterfowl. It forages by low fast flight, quartering and high soaring, or still-hunting from a perch. It seizes or strikes prey in flight by a stoop or direct flying attack that becomes a vigorous chase. It also hawks flying insects or glides from a perch to take prey on the ground. It sometimes hunts cooperatively in pairs. It follows grass-fires, farm machinery, livestock, shooters, and other raptors to seize flushed prey, and it robs other raptors.

BEHAVIOUR

The Black Falcon is usually observed soaring lazily or perched on the top of a dead tree, or more rarely on a fence post (although not on wires like the Brown Falcon). It probably spends much of the day on the wing. Sometimes it can be seen harassing other raptors vigorously. In the breeding season, solitary birds or pairs soar and perform aerobatics with exaggerated wing-beats, diving, and mock attack-and-parry, or fast chases at low level. Courtship feeding occurs. Some regular migratory movement apparently occurs northwards from south-eastern Australia for the winter and south for the summer, as well as nomadism to local abundances of quail or to plagues of other species of prey.

BREEDING

The laying season is May to November, usually July to September. Pairs nest solitarily. The Black Falcon uses the large stick nest of another raptor or corvid, in the top of a living or dead tree or (rarely) on an electricity pylon, 4–14 m above ground. The clutch size is usually three or four eggs, ranging from one to five. Incubation takes about 34 days, and the nestling period is 38–49 days. The period of dependence after fledging is poorly known but lasts at least two weeks and probably longer. The oldest recovered banded bird was 12 years; longevity in captivity is up to 20 years.

THREATS AND CONSERVATION

The endemic Black Falcon is not nationally threatened. It is generally uncommon but widespread; it can be locally common in the arid zone in wet years, although it is usually scarce there. The population fluctuates about an apparently stable average, or it might have increased in agricultural areas, where it benefits from increased abundance of prey, including introduced species. The thickness of its eggshells was not significantly reduced by DDT use in Australia.

Peregrine Falcon
Falco peregrinus

DESCRIPTION

The specific name ('wandering') alludes to the migratory habits of Peregrines of the Northern Hemisphere, although Australian birds are sedentary as breeding adults. The species is characterised by short tarsi, large feet, and long toes with reticulate tarsal scalation.

The Peregrine Falcon is 36–50 cm long (tail less than half), with a wingspan of 81–106 cm. The average male weighs 552 g, females 823 g. It is similar in size to the Brown and Black Falcons but with shorter wings and tail; it is stockier than other Australian falcons. It is a compact, heavily built falcon with a full black helmet, deep chest, broad-based wings, and broad rump and tail. It is swift and powerful in flight, appearing dark.

The **adult** has a black head and cheeks, forming a full helmet. The upperparts are slaty blue-grey, somewhat paler on the rump, with black wing tips and finely barred wings and tail. The underparts are white (conspicuously so on the chest), with a grey wash and fine dark bars on the abdomen. The female is somewhat browner than the male, with a stronger buff suffusion on the underparts. Some birds are paler dorsally and more rufous ventrally, most commonly in south-western Australia. The cere and eye-ring are bright yellow, the eyes brown, and the feet bright yellow. The **juvenile** is darker and browner than the adult. The upperparts have brown feather edging, and the underparts are deep buff with black streaks on the breast and coarse wavy bars on the thighs and flanks. Inland birds tend to be paler. The cere and eye-ring are pale blue at fledging to pale yellow when older, the eyes brown, and the feet dull yellow. The **chick** has white down.

The Peregrine Falcon is a solitary, aggressive falcon occurring in most habitats but characteristic of cliffs, escarpments, and wetlands. Its flight action is rapid and powerful with deep, rhythmic wingbeats. It glides on flat wings. It soars with the wings held stiffly out from the body, the trailing edges straight, and the pointed tips sometimes slightly upswept (see figures 10.11 and 10.12). In flight it appears somewhat front-heavy with large head and bill, deep chest, and a short tail. It flies and soars strongly at great heights and stoops with closed wings and compact, bullet shape. It can seize or strike down quite large birds and readily feeds on the ground. At rest it has a stocky, heavy-shouldered appearance, showing a pale bib and large feet. Its most commonly heard call is a hoarse, screaming chatter. It also utters clucking and whining calls.

Figure 10.10a Soaring and gliding

Figure 10.10b Soaring and gliding

(a)   (b)

Figure 10.11 Soaring (a) Peregrine Falcon (b) Australian Hobby

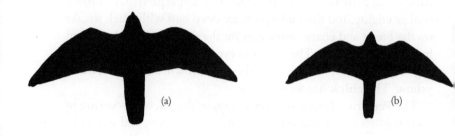

Figure 10.12 Gliding (a) Peregrine Falcon (b) Australian Hobby

DISTRIBUTION

The cosmopolitan Peregrine Falcon occurs naturally throughout mainland Australia and Tasmania, where it has had time to differentiate into an endemic race (subspecies *macropus* meaning 'large foot').

FOOD AND HUNTING

The Peregrine Falcon eats mostly flocking birds, particularly pigeons, parrots, and starlings, but also larger species up to the size of herons, ibises, and waterfowl; it commonly takes seabirds on the coast. It occasionally eats large insects and rarely fish, reptiles, small mammals or carrion. Its hunting is diurnal, crepuscular, and sometimes nocturnal by artificial light. It forages by still-hunting from a high perch, by high quartering and soaring, or by low fast flight. It seizes or strikes prey in flight by a long, shallow or slanting stoop, or by a direct flying attack that can become a vigorous chase. It also hawks flying insects, sometimes hunts cooperatively in pairs, and flushes prey from cover.

BEHAVIOUR

The Peregrine Falcon is typically seen in flight: often soaring, flying directly and purposefully at moderate heights, stooping, or sometimes dashing past at low level with swishing wings. It also perches on prominences, such as cliffs, or on structures such as towers and buildings. Its flight appears masterful and controlled, even in windy conditions. Peregrines soar and perform diving and other aerobatics, and often harass larger birds (see figure 10.13). Members

Figure 10.13 Aerobatic V flight in display

of a pair behave similarly in display flights, which culminate in courtship feeding (with some aerial food passes) and mutual inspection of nesting sites, with Bowing Displays and calling.

BREEDING

The laying season is August to October. Pairs nest solitarily. The Peregrine Falcon uses a scrape on a cliff ledge or quarry face, the old stick nest of another raptor in a tree or on an electricity pylon, a tree hollow, or the ledge of a structure, such as a building, dam wall or silo. Height ranges from below ground in mine shafts to 150 m above ground. The clutch size is usually three or four eggs, ranging from one to five. Incubation takes 33 days, and the nestling period is 38–40 days. In south-eastern Australia 58% of nests were successful: 1.23 young fledged per territory, 1.4 per pair, 1.83 per clutch started and 2.16 per successful nest around Canberra; 1.3–1.57 young per pair, 2.0 per successful pair and 1.3–1.55 per clutch started in Victoria; 64% nest success in Tasmania. The period of dependence after fledging lasts up to two or three months, with young sometimes remaining in the area of the nest for up to eight months, after which they disperse widely (up to 500 km has been recorded, although usually about 60 km for males and 130 km for females). Age at first breeding is 2 years for females and 3 years for males.

THREATS AND CONSERVATION

The Peregrine Falcon is not globally or nationally threatened. In Australia it has increased in agricultural and urban areas with the increase in suitable food and its use of city buildings for nesting sites. The thickness of its eggshells was reduced by DDT use in Australia, probably enough to cause local declines in breeding success and disruption of local gene pools in the south-east. Eggshell thickness has returned to normal following the ban on the use of DDT. In some areas the Peregrine is heavily (but illegally) persecuted by pigeon-fanciers. Human disturbance at nest sites and the illegal taking of eggs and nestlings can also cause local reduction in breeding success.

Nankeen Kestrel (Australian Kestrel)
Falco cenchroides

DESCRIPTION

The specific name alludes to the bird's similarity to the Common Kestrel (*Falco tinnunculus*) of the Old World. The species is characterised by moderately long tarsi, with coarsely reticulate scalation, and short toes.

The Nankeen Kestrel is 30–35 cm long (tail about half), with a wingspan of 66–78 cm. The average weight of males is 163 g, females 173 g. It is the smallest Australian raptor (other than the male Collared Sparrowhawk): smaller and slighter than the Black-shouldered Kite, with narrower wings and longer, rounded tail. It is a small, delicate falcon with distinctive rufous upperparts and pale underparts, black wing tips, and a band near the tail tip.

The **adult male** has a grey (or sometimes rufous) head with fine dark streaks and an indistinct malar stripe. The upperparts are rufous with sparse black streaks or spots and black wing tips. The rump and tail are grey, with a black subterminal tail band and white tip to the tail. The underparts are buff, more rufous on the chest and flanks, with fine dark shaft streaks. The underwings are pale and faintly barred. The cere and eye-ring are yellow, the eyes brown, and the feet yellow. The **adult female**'s head, rump and tail are rufous; the upperparts and underparts tend to be more heavily marked with black, and the tail often has some barring. The rump might be grey, and the tail might be lightly washed grey. The **juvenile** is similar to the adult female. The upperparts tend to be more heavily streaked with black, although this character varies. The tail has more and narrower bars. **Fledglings** have a richer rufous rump and tail and a more conspicuous pale trailing edge to the wings in flight. The cere and eye-ring are pale grey with a green or yellow tinge, changing to yellow as the bird gets older, the eyes are brown, and the feet yellow. The **chick** has white down.

The Nankeen Kestrel is a solitary or loosely gregarious falcon of most open habitats, characteristic of farmland with scattered trees and of inland shrublands and woodlands. Its flight action is rapid and winnowing, with sweeping glides on flat wings (see figure 10.14);

Figure 10.14 Soaring and gliding

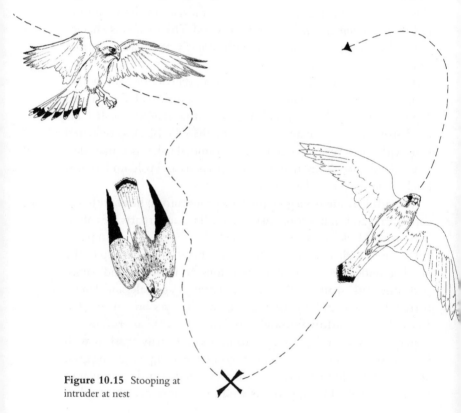

Figure 10.15 Stooping at intruder at nest

it hovers with the body horizontal or hangs in the wind with motionless wings flexed and held above the back; it dives steeply on to ground-dwelling prey with closed wings (compare with the Black-shouldered Kite). Its most commonly heard call is a strident, rapid, shrill chatter (three notes per second); also a short single note repeated (one note per second) and upslurred chittering.

The Nankeen Kestrel is slimmer, longer-tailed, and buffier from below than the Black-shouldered Kite; it also has a black subterminal tail band and lacks the Kite's underwing markings. Pale Brown Falcons can be confused at rest but are larger with a double cheek-mark and long legs. In a poor view the Kestrel can be confused with the Australian Hobby, which is darker with more pointed wings and a shorter, squarer tail.

DISTRIBUTION

The Nankeen Kestrel occurs throughout Australia, although rarely in Tasmania. It is resident on offshore islands, including Norfolk Island and Christmas Island. Migrating individuals reach New Guinea, where there is also an endemic race (*baru*) in the montane grasslands, and other islands to the north as far as Java. Vagrants reach New Zealand.

FOOD AND HUNTING

The Nankeen Kestrel eats mostly invertebrates, particularly insects such as grasshoppers and crickets; also small mammals (mice), birds up to sparrow and starling size, and reptiles (mainly skinks and small dragons), especially when breeding. It forages by high quartering and hovering, or still-hunting from a perch. It seizes prey on the ground by a dive or glide. It also hawks flying insects, but rarely chases small birds.

BEHAVIOUR

The Nankeen Kestrel is a common roadside raptor, typically seen perched on fences, wires, utility poles, and dead trees or shrubs, or hovering over paddocks, although it is also common in treeless inland areas away from roads. In settled areas it commonly perches on buildings. Solitary birds perform a winnowing display flight with bursts of short, rapid beats and glides on drooped wings, accompanied by chattering. Sometimes the body is canted to produce a flash pattern. Members of a pair soar and perform mock attack-and-parry manoeuvres; they also greet at perches with a Bowing Display and occasionally allopreen. Food passes are sometimes aerial, although usually on a perch near the nest. Kestrels defend their nests by swooping and striking at intruders (see figure 10.5).

BREEDING

The laying season is August to December in Australia. Pairs usually nest solitarily, although sometimes semi-colonially in concentrated breeding habitat amid extensive hunting habitat, such as woodlots in farmland. The Kestrel nests in a variety of sites: from sink-holes and mine shafts below ground to tree hollows, old nests of other birds, cliffs, ledges on buildings, or machinery, 2–31 m above ground. Most commonly it nests in tree hollows and on cliffs. The eggs are laid in a depression or scrape. The clutch size is usually three to five eggs, ranging from one to six. Incubation takes 28 days, and the nestling period is 31–35 days. Breeding success has been measured as 68% nest success (of attempts), 66% hatching success (of eggs laid), 45% fledging success (of eggs laid), and 1.26 young fledged per attempt. The period of dependence after fledging lasts up to two months, after which young disperse or migrate widely (up to 800 km has been recorded, although usually within 10 km of the banding site). Young are sexually mature and sometimes breed when a year old.

THREATS AND CONSERVATION

The Australian Kestrel is not globally or nationally threatened. It is widespread and abundant throughout Australia, where it might have increased in numbers; it has benefited from clearing and the introduction of suitable prey. It has recently colonised Lord Howe, Norfolk and Christmas Islands and might be colonising New Zealand. The thickness of its eggshells was not significantly reduced by DDT use in Australia.

Chapter 11

Threats, conservation and the future

Most Australian raptors face threats of one sort or another, commonly more than one acting in concert, and all such threats arise ultimately from prevailing human attitudes to land, water, and wildlife. Only the most abundant and adaptable species, such as the Black-shouldered Kite, Black Kite, Brown Falcon and Nankeen Kestrel, give no cause for concern at present. However, if experience in Europe or North America is a warning (and it should be), then the situation even for common species could change in the near future.

At present the most threatened Australian raptors are our special endemic species, which occur nowhere else in the world: the Square-tailed Kite, Red Goshawk, and Grey Falcon. The next most threatened species, of national concern although not listed in one of the 'threatened' categories, are also endemics: the Letter-winged Kite and Black-breasted Buzzard. The Buzzard is still common in northern Australia, whereas the Letter-winged Kite probably deserves greater concern than it currently receives. Several other species might deserve greater official recognition as of special concern, if not yet threatened: the scarce Black Falcon and the Pacific Baza and Grey Goshawk, which have restricted ranges.

HUMAN IMPACT ON HABITAT

Destruction, degradation or disturbance of habitat, whether of nesting sites or foraging habitat, is identified as a threat to most Australian raptors. 'Habitat loss' refers to deforestation in humid regions, clearing of semi-arid woodlands for marginal agriculture, overgrazing of arid and semi-arid rangelands, drainage of wetlands, and impact on estuarine and littoral habitats. It is the major factor in the plight of our threatened and near-threatened species, and has reduced the regional populations of many others. Destruction of native vegetation is clearly the single most important threat to Australia's raptor communities and indeed to wildlife communities in general, not to mention a key element in the degradation of the land and water resources that sustain humanity. Yet vegetation clearance proceeds apace, with the likelihood that it will accelerate in the tropics as we devise more efficient ways of removing trees from large expanses of the landscape.

Timber harvesting is often seen as a threat to forest-dependent raptors, particularly where logging occurs on a broadscale clear-felling pattern and the next harvest occurs before the regrowth reaches full maturity. This is the case in Tasmania for the Grey Goshawk and for the endemic race of the Wedge-tailed Eagle, which is sensitive to disturbance around the nest. However, the indications are that most Australian forest raptors are tolerant of logging on a selective or sustained-yield basis where there is some canopy retention during cutting operations, adjacent blocks are not harvested simultaneously, and reserves and corridors exist across the logging mosaic. Only the Red Goshawk requires old-growth forest characteristics, in lowland fertile forest, as nesting habitat. Forestry is therefore not such a major issue compared with habitat clearance on other lands, yet the focus and energies of the environmental movement do not reflect this.

PESTICIDES AND POLLUTION

DDT reduced the eggshell thickness of several aquatic and bird-eating raptors in Australia, probably enough to cause local breeding failure and population disruption in some areas, although without the spectacular population crashes of the northern hemisphere. The Grey Falcon was one species affected in agricultural areas. The Red Goshawk was probably also affected in some areas, although there are insufficient egg samples to confirm this suggestion. The eggshell thickness of Australian Peregrine Falcons has returned to normal since the use of DDT was banned, and presumably the same applies to the other affected species, with no lasting consequences for population levels. However, a legacy could be the loss of the genetic distinctness of local Peregrine populations: formerly, the cliff-nesting and stick-nesting Peregrines had recognisably different eggs from riverine hollow-nesting Peregrines. Furthermore, DDT is still used in New Guinea and other islands to the north, where migratory Australian raptors or their prey might be ingesting pesticides and developing pesticide loads.

Dieldrin, another notorious organochlorine pesticide, could have caused direct raptor mortalities, but any effects on populations are unknown and presumably have now been reversed, as suggested by the recovery of the Osprey and Peregrine Falcon. Like DDT, dieldrin was banned from agricultural use in Australia because residues in beef affected export prospects, not because of any official concern for its impact on wildlife.

Secondary poisoning from rodenticides and organophosphate insecticides is likely to be underestimated as a cause of raptor mortality in Australia. Raptors have ample opportunity to ingest poisoned rodents, birds or locusts during broadscale baiting or spraying attempts to control plagues, and raptors have died *en masse* in Africa during pest-control programs. Nevertheless, it is mainly the common Australian species that are affected. Secondary poisoning from rabbit baits containing 1080 (sodium monofluoroacetate) or pindone might also occur. Widescale baiting for dingoes and foxes in vast inland areas might significantly affect regional populations of scavenging raptors, but data are lacking. We know a little about the lethal effects of some of these poisons fed directly to captive raptors, but next to nothing of their impact under field conditions on populations of raptors.

Aquatic raptors, in particular, are likely to be carrying a mixture of chemical pollutants acquired via the food chain from waterways affected by urban, industrial, and agricultural discharge or run-off, as illustrated by the effect of DDT on the eggshell thickness of most Australian aquatic species. Chemicals with demonstrated toxic effects on raptors or their breeding success in other countries include polychlorinated biphenyls (PCBs) and heavy metals, such as mercury. At present there are no known population effects on Australian raptors, but the risk remains while we, in the driest continent on Earth, treat our precious waterways as open sewers. Some raptors might ingest lead shot in the tissues of their prey (notably waterfowl) or in carrion, and so suffer lead poisoning. Lead, entering food chains from the air pollution in our towns and cities, could also contaminate individual raptors that live in the urban environment.

DIRECT PERSECUTION

Raptors are killed illegally by shooting, trapping, poisoning, and other inhumane methods for perceived or alleged damage to livestock and domestic animals, and occasionally for the clandestine trade in stuffed specimens. The species most persecuted are the Wedge-tailed Eagle in sheep-raising regions, the Brown Goshawk around poultry, pigeon lofts, and aviaries, and the Peregrine Falcon by pigeon-fanciers, although few species are immune to destruction on suspicion. In past decades persecution of eagles was intense, but knowledge, understanding, and legal protection seem to be prevailing, although the killing and calls for removal of legal protection continue. These days killing of eagles is being upstaged by the persecution of Peregrine Falcons by pigeon-fanciers. These people wage a campaign against the Peregrine based largely on untruths, the most scurrilous of which is that the Peregrine is an introduced species bred in captivity and released in large numbers by wildlife agencies.

After a number of prosecutions, egg-collecting seems to be declining, although there are still underground operators with their secret networks and closet hoards of empty shells. Historically, egg collection might have lowered breeding success in local populations of uncommon or rare species (which are the most sought-after). However, without the historical egg collections we would not have proof of the effects of DDT on our raptors. Egg-collecting has long since ceased to serve any useful purpose. Raptor enthusiasts should refrain from activities or disclosures that might enable anyone other than their most trusted colleagues to find raptor nests.

Clandestine falconers or would-be falconers operate illegally in Australia, as revealed by the trickle of birds turning up with jesses on their legs and often showing signs of mishandling or behaviour problems. Taking small numbers of wild raptors for falconry is unlikely to affect populations, but the behaviour of escaped birds can bring raptors into disrepute. Of more concern is the potential for raptors to be smuggled to overseas falconers, as illustrated by two recent cases. One case involved a group of Arabs who were under surveillance while they travelled Australia with falconry equipment and raptor books. Some of the party left the country clean via an official airport, but others in the party left the country undetected, probably in a private jet from a remote airstrip (with who knows how many raptors?). Another case involved a person who learnt raptor trapping and banding skills legitimately, by assisting an approved research project, then misused the knowledge and skills to trap a pair of Red Goshawks, which were then smuggled out of Australia, probably to the Middle East. Perhaps trafficking is one reason for continuing theft of eggs. The lesson for raptor students is: be very careful whom you trust with knowledge of nests!

RESEARCH

It is a truism that effective conservation depends on sound knowledge of the biology and ecological requirements of the subject species, in order to institute appropriate management strategies. It follows, then, that we should know or find out all we need to know about conserving our most threatened raptors. Those species are, of course, our unique, taxonomically distinct endemics, some of which are in Australian or regionally endemic genera (some monotypic). Yet the level and direction of research do not reflect this need.

Wildlife agencies have conducted major ecological studies on four species (Wedge-tailed Eagle, Peregrine Falcon and Red Goshawk, with White-bellied Sea-Eagle unpublished), and limited research or monitoring has been carried out by several state wildlife departments (Osprey, Peregrine, Red Goshawk). Tasmania is unique in having an official raptor research and survey program, largely because of the interests of key personnel in the wildlife department but perhaps also because the populations of significant species there are measured in mere tens of breeding pairs. The CSIRO has studied Whistling Kites and Sea-eagles in the tropics; nothing substantial has been published as yet. Otherwise, the best ecological research on Australian raptors has been or is being conducted by postgraduate students, on common species, with limited continuation at higher academic levels (and then mostly on the Peregrine, although the Letter-winged Kite is receiving attention).

The species that have missed out on high-quality ecological research are the Square-tailed Kite, Black-breasted Buzzard, Pacific Baza, and Grey Falcon. The greatest need is for population surveys and research into the biology and ecology of the Red Goshawk (in eastern Australia), Square-tailed Kite, and Grey Falcon, with the Black-breasted Buzzard and perhaps Black Falcon next in priority. After the wave of raptor research in the 1980s, culminating in *HANZAB* 2, the current trough is, or should be, alarming.

RESERVES

Another truism is that national parks and other conservation reserves in Australia, particularly in the heavily settled south, are delineated by soil-type boundaries. With few exceptions, nature reserves are the 'useless', rugged, infertile land that was not wanted for anything else. Such areas have high wilderness and scenic values, but relatively poor biodiversity values when compared with the richer habitats on flat, fertile land that people want to clear, log, drain, or build on. Yet the focus and energies of the conservation movement do not reflect the greatest need: we have opted for the conservation of wilderness monuments rather than the conservation of biodiversity, while the extinction of our wildlife species continues.

Reserves have contributed to raptor conservation incidentally by protecting populations of common species and a few pairs of significant species, but in the most disturbed regions they are rarely large enough to preserve viable raptor populations within their boundaries. Therefore, although they have value, existing reserves generally have not preserved the best raptor habitats. A system of new reserves targeting remaining habitats with high biodiversity value, and based on biological criteria, is needed. If raptors were used as flagship and indicator species for ecosystem conservation, a regional system of reserves conserving viable populations of significant raptor species would also automatically encompass other important environmental values. This is because raptors are at the top of the food chain and require large areas to support them.

However, reserves alone cannot be relied on to conserve raptors in the future. The fate of habitats on land outside reserve boundaries is likely to be more important, as reserves become islands and threatening processes continue around them. On disturbed lands, buffer zones and other protection measures around active nests, such as restricting disturbance during the breeding season, are effective.

HABITAT RESTORATION

One of the best things that humans can do for raptors that depend on forest and woodland is to regenerate some of the indigenous tree cover on our battered and derelict farmland. This would help to restore many other environmental values, not least the economic productivity of areas that are over-cleared and suffering from various forms of land degradation. Rivers and streams should receive particular attention, because they are important to raptors and other wildlife. A national scheme to win back the waterways would have far-reaching benefits for the human population as well as wildlife. Elements of such a scheme should include fencing-off creeks and rivers to exclude stock and to allow the regeneration of the native riparian vegetation.

Many more shelter belts and woodlots of locally indigenous trees throughout our agricultural and urban lands, properly constructed as functioning windbreaks with appropriate profiles, and including a shrub layer, would be of great benefit. The Landcare movement gives cause for hope.

Focused habitat enhancement, such as provision of nesting sites, can help, particularly where natural nesting sites have been lost from the landscape. Ospreys have nested on artificial platforms, Kestrels have bred in nest-boxes, and Peregrines have bred on suitably prepared ledges on city buildings. Such hands-on approaches are popular overseas where habitat destruction has left little option, but in Australia we should prevent loss of natural nesting sites in the first place, particularly as our most threatened species are unlikely to accept artificial sites. There is also a danger that, for instance, the willingness of Ospreys to nest on platforms will be used as an excuse for allowing their tree sites to be destroyed.

PEST MANAGEMENT

Integrated pest management is a concept that seeks to control agricultural pests with minimal use of hazardous chemicals. The approach is to understand the ecology of the pest and its natural enemies, in order to use pesticides sparingly and strategically at the most appropriate stage of the pest's lifecycle while taking full advantage of natural and biological controls. The benefits to raptors are obvious, but raptors could also have a positive role in the pest control operation, for instance, through human provision of hunting perches in areas affected by pests. Farmers would benefit, too, through savings on expensive chemicals that would not be needed in such large amounts, as well as a reduced chance of pest resistance, and hence greater crop yields. The concept seems disappointingly slow to catch on.

Similarly, there are raptor-friendly rodenticides that carry a low risk of secondary poisoning, if used strictly according to instructions in an integrated operation that combines removal of the pest's shelter, trapping, collection and disposal of poisoned rodents, and so forth. The most notable is coumatetralyl, sold as Racumin by Bayer. It is widely used and promoted in Africa and elsewhere in the world as safe, effective and resistance-free, yet in Australia we persist in using the more toxic forms of anticoagulant, such as brodifacoum.

EDUCATION

Some pastoral interests still believe and advocate that the Wedge-tailed Eagle is a significant economic threat to the sheep industry. Similarly, the pigeon lobby's outbursts against Peregrines are based on ignorance or misinformation. These beliefs show that the results of research are not reaching those most in need of the knowledge. The need for public education about raptors is therefore obvious, but it seems that the only long-term hope is with the young.

I have few ready answers, but perhaps a two-pronged educational approach is needed from early school age and using the most influential media to bring the subject into people's homes as well as school curricula (TV, video, and perhaps computer packages!). First, a general appreciation of ecological principles and the role of predators, with accurate presentation of natural history on Australian raptors, is needed. Second, a focused campaign to correct public misconceptions, arising from the misinformation propounded by the anti-eagle and pigeon lobbies, is also needed. The *Hunters of the Skies* TV series was a great start; we need more documentaries on nature and natural processes, incorporating raptors as part of a whole and using them as a catchy way to present some of the principles of ecology and conservation. Meanwhile, articles in nature or geographical magazines have a positive impact, but I suspect that they largely preach to the converted. Raptors could do with more positive exposure in everyday reading media, such as newspapers and magazines.

Research is one thing; getting it to the public in digestible form is entirely another, and ultimately of real benefit to raptors. Bird and raptor enthusiasts could take every opportunity, via letters to the editors of newspapers, to counter the anti-raptor propaganda by presenting the facts. Perhaps a few letters in the direction of biased TV reporters wouldn't hurt, either.

REHABILITATION

There is a disturbing trend for rehabilitation and captive breeding to be promoted as the panacea for the ills of raptors, with the number of rehabilitators growing rapidly. The energies and resources of raptor enthusiasts and wildlife administrators are focused on rehabilitation at the expense of research. Claims are made in the media or influential publications that captive breeding will save the Peregrine Falcon in Australia, and funds are raised for rehabilitation centres with the claim that patching up a few individuals and returning them to the wild will save the Wedge-tailed Eagle. Yet these species do not need 'saving'; many others are more deserving. Such claims damage the credibility of raptor enthusiasts and could hinder attempts to conserve species that are genuinely threatened. The role and benefits of raptor rehabilitation in Australia need to be seen in perspective.

Most raptor cases brought to rehabilitation centres for treatment have been harmed by human agency; therefore rehabilitation makes up, to a slight degree, for the human impact on raptors. However, Australian raptor numbers are such that rehabilitation does not make a significant difference to wild population levels. The real benefits of rehabilitation are in education, public relations, knowledge, and the perfection of techniques on common species for future use on rare species if necessary. Involvement of the public, face to face with rehabilitated birds, is probably the most powerful educational tool we have if such exposure is conducted properly with sufficient coverage of the real issues.

Much the same could be said of captive breeding and release of offspring, of which there is very little in Australia at present, although these activities might increase in popularity. Releases would not significantly increase the wild population of any Australian raptor species, at least until we know what is limiting the population of, say, the Red Goshawk or Grey Falcon, and we have reversed the processes that threaten them. Releasing birds is pointless if there is no vacant habitat in which they can live. However, captive breeding does enable crippled birds to contribute their genes to wild populations, and it provides knowledge and skills. Captive breeding should be undertaken only under scientific supervision, using unreleasable crippled birds as parent stock, if our rarest species are to benefit from it. Present indications are that taking viable eggs or birds from the wild would not be outweighed by subsequent population increases from release of captive offspring.

THE FUTURE

An optimistic scenario is that over the next few years humanity will progress to a more sensitive role in the Australian environment: that of steward. We will repair our land and waterways, stabilise our population, and reduce our consumption and pollution. We will manage our farmlands to maintain functioning ecosystems on them, and adopt the principles of integrated pest control. We will have a system of true biosphere reserves dedicated for their high biodiversity values. We will understand and appreciate our raptors, conduct research on our most threatened species, and put appropriate management strategies in place to conserve them. The alternative scenario is to carry on much as we are, degrading our environment, and watch the disappearance of many of our raptors. The loss will start with our unique Australian endemics—the present generation has already effectively lost the Red Goshawk from New South Wales.

Glossary

allopreen Preen (groom) another bird (in raptors, usually the partner's head).

alula The small, pointed feathers attached to the 'thumb' of the wing, beyond the carpal joint.

anticoagulant A poison that works by preventing blood from clotting, and thus causing internal bleeding.

bigamous Having two mates at a time.

biodiversity Biological diversity or richness: the total number of plant and animal species living in an area.

cache To hide surplus food, for example in crevices, for later retrieval.

carpal The joint at the bend of the wing, corresponding with the wrist in humans.

cere The soft, fleshy skin at the base of the bill, in which the nostrils of raptors are situated.

convergence Evolution, by unrelated animals, towards a similar physical form.

corvid A member of the family Corvidae; in Australia crows and ravens (genus *Corvus*).

coverts Feathers overlying a specified part of a bird's anatomy, for example, secondary coverts.

crepuscular Active at dawn and dusk.

dichromatic Two plumage types within a species, for example, sexual dichromatism: male and female are coloured differently.

dihedral The formation of the wings in gliding flight when the wings are held raised above the horizontal plane, in a shallow V.

dimorphic Having two physical forms, for example plumage dimorphism: two life-long plumage types within a species, not related to age or gender; sexual dimorphism: male and female of different size.

diurnal Active by day.

emarginated Having narrow inner and outer webs on the ends of the primary feathers, thus producing separated feathers and slotted tips on the fully spread wings.

endemic Unique to a particular region.

extralimital Outside a species' normal distribution.

fledging First flight from the nest.

flight feathers The remiges (primary and secondary wing feathers).

gliding Passive forward or downward, non-flapping flight on extended wings.

Gondwana The ancient southern supercontinent, consisting of Africa, Madagascar, India, Australia, South America, and Antarctica, before they were separated by continental drift.

hovering Maintaining a stationary position in relation to the ground with active flapping flight head to wind, by matching forward velocity to wind velocity.

immature A plumage stage between juvenile and fully adult plumages.

indigenous Occurring naturally in a specified region.

irruption Mass movement of a species to an area.

juvenile The first plumage on leaving the nest.

lores The area between the eyes and the bill.

malar stripe A dark streak extending below the eye.

mantle (1) The area between the hindneck, back and scapulars; (2) to spread the wings in threat or to cover food protectively.

megapodes Mound-building, fowl-like birds (Megapodiidae), including the Australian Brush-turkey and Malleefowl.

melanistic A dark colour form, either black or blackish.

metapopulation A regional population consisting of several local, discrete subpopulations whose members can disperse among themselves.

monogamous Having one mate.

monotypic Having one member: a monotypic genus contains a single species; a monotypic species has no geographical races (subspecies).

morph A life-long plumage type, unrelated to age or gender.

nape The back of the neck.

nocturnal Active by night.

occipital Referring to the back of the head, where it joins the nape.

orbital The skin surrounding the eye.

passerine A bird of the order Passeriformes or perching birds: the 'bush birds' or songbirds with simple bills and perching feet having three toes forward and one back.

piscivorous Eats fish.

poising Maintaining a stationary position in relation to the ground by wind-hanging without active flapping.

polyandrous A female mated to more than one male.

polygamous Having more than one mate.

polygynous A male mated to more than one female.

polymorphic Having more than one plumage morph.

primaries The long flight feathers on the outer half of the wing, attached to the 'hand'.

quartering Repeatedly searching an area thoroughly, in flight.

radiation Evolutionary divergence of related species into a variety of physical forms.

rectrix (retrices) The main tail feather(s).

refugium place of refuge; an area in which a population can survive through a period of unfavourable conditions.

remex (remiges) The main wing feather(s) (primaries and secondaries).

reticulate A mosaic pattern of small scales on the tarsi.

rodenticide A poison for killing rodents (rats and mice).

scapulars The feathers on the sides of the back, attached to the shoulder blades, and overlying the inner parts of the wings.

scutellate Large, plate-like or shield-like, overlapping scales on the tarsi.

secondaries The short flight feathers on the inner half of the wing, attached to the 'arm'.

secondary poisoning Intoxication by eating an animal that was poisoned and thus ingesting the poison in that animal's tissues.

soaring Gaining height in non-flapping flight, on fully spread wings, by riding thermal currents or updraughts.

spicules Small, spiny projections on the undersides of the feet.

stooping A diving attack with closed wings, in which contact with flying prey is made in the air.

tarsus The lower half of the visible part of a bird's leg, below the ankle joint. The true knee is hidden by flank feathers.

taxonomy The study of the evolutionary and genetic relationships between organisms.

tomial tooth A projection on the cutting edge of the upper mandible.

transect Traversing an area on a straight course.

Wallacea The Indonesian islands east of Wallace's Line, and thus part of the Australasian faunal region: Sulawesi, the Moluccas, and the Lesser Sunda Islands.

Bibliography

Only the most recent papers, published concurrently with or since *HANZAB 2*, are listed here. Earlier ones are listed in *HANZAB*, and in Penny Olsen's book *Australian Birds of Prey* (New South Wales University Press, 1995).

Aumann, T. 1993. Seasonal movements of the Brown Goshawk *Accipiter fasciatus* in Australia. In Olsen (ed.) (1993), pp. 228–242.

Baker-Gabb, D. 1993a. Wing tags, winter ranges and movements of Swamp Harriers *Circus approximans* in south-eastern Australia. In Olsen (ed.) (1993), pp. 248–261.

Baker-Gabb, D. 1993b. Auditory location of prey by three Australian raptors. In Olsen (ed.) (1993), pp. 295–298.

Baker-Gabb, D.J. 1994. Threatened raptors of Australia's tropical forests. In Meyburg & Chancellor (eds) (1994), pp. 241–244.

Beehler, B.M., Pratt, T.K. & Zimmerman, D.A. 1986. *Birds of New Guinea*. Princeton University Press, Princeton, NJ.

Beruldsen, G. 1995. *Raptor Identification*. Author, Brisbane.

Bird, D.M., Varland, D.E. & Negro, J.J. (eds). 1996. *Raptors in Human Landscapes*. Academic Press, London.

Brereton, R.N. & Mooney, N.J. 1994. Conservation of the nesting habitat of the Grey Goshawk (*Accipiter novaehollandiae*) in Tasmanian state forests. *Tasforests* 6: 79–91.

Bretagnolle, V. & Thibault, J.C. 1993. Communicative behaviour in breeding Ospreys (*Pandion haliaetus*): Description and relationship of signals to life history. *Auk* 110: 736–751.

Brickhill, J. 1993. Abundance of raptors in the Riverina, 1978–1987. In Olsen (ed.) (1993), pp. 262–272.

Britton, P.L., Britton, H.A. & Rose, A.B. 1996. Note on the diet of the Letter-winged Kite near Charters Towers, north Queensland. *Sunbird* 26: 63–65.

Brooker, M. 1996. Morphometrics of the Wedge-tailed Eagle *Aquila audax*. *Corella* 20: 129–135.

Burnett, S., Winter, J. & Russell, R. 1996. Successful foraging by the Wedge-tailed Eagle *Aquila audax* in tropical rainforest in north Queensland. *Emu* 96: 277–280.

Burton, A.M. & Alford, R.A. 1994. Morphometric comparison of two sympatric goshawks in the Australian wet tropics. *Journal of Zoology (London)* 232: 525–538.

Burton, A.M., Alford, R.A. & Young, J. 1994. Reproductive parameters of the Grey Goshawk (*Accipiter novaehollandiae*) and Brown Goshawk (*Accipiter fasciatus*) at Abergowrie, northern Queensland, Australia. *Journal of Zoology (London)* 232: 347–363.

Burton, A.M. & Olsen, P. 1997a. Niche partitioning by two sympatric goshawks in the Australian Wet Tropics: Breeding-season diet. *Wildlife Research* 24: 45–52.

Burton, A.M. & Olsen, P. 1997b. A note on inter- and intraspecific differences in the diets of sympatric Brown Goshawks and Grey Goshawks in the non-breeding season. *Australian Bird Watcher* 7:138–141.

Burton, A.M. & Olsen, P. 1997c. A note on hunting behaviour in two sympatric goshawks in the Australian Wet Tropics. *Australian Bird Watcher* 7:126–129.

Bustamante, J. 1995. The duration of the post-fledging dependence period of Ospreys *Pandion haliaetus* at Loch Garten, Scotland. *Bird Study* 42:31–36.

Cameron, M. & Olsen, P. 1993. Significance of caching in *Falco*: Evidence from a nesting pair of Peregrine Falcons *Falco peregrinus*. In Olsen (ed.) (1993), pp. 43–54.

Carlier, P. 1994. Influence of male–female relationships on parental behaviour in two pairs of Peregrine Falcons *Falco peregrinus*. *Butlleti del Grup Catala d'Anellament* 11: 75–82.

Carlier, P. 1995. Vocal communication in Peregrine Falcons *Falco peregrinus* during breeding. *Ibis* 137: 582–585.

Carlier, P. & Gallo, A. 1995. What motivates the food-bringing behaviour of Peregrine Falcon throughout breeding? *Behavioural Processes* 33: 247–256.

Clancy, G.P. 1993. The conservation status of the Osprey *Pandion haliaetus* in New South Wales. In Olsen (ed.) (1993), pp. 192–195.

Coates, B.J. 1985. *Birds of Papua New Guinea*, volume 1, *Non-Passerines*. Dove, Brisbane.

Collar, N.J., Crosby, M.J. & Stattersfield, A.J. 1994. *Birds to Watch 2. The World List of Threatened Birds*. BirdLife International, Cambridge, UK.

Cupper, J. & Cupper, L. 1981. *Hawks in Focus*. Jaclin, Mildura.

Currie, L., Klapste, J. & Baker-Gabb, D. 1993. A preliminary study of longevity in Brown Goshawks wintering at Werribee, Victoria. In Olsen (ed.) (1993), pp. 243–247.

Czechura, G.V. 1993. The Pacific Baza *Aviceda subcristata* in south-eastern Queensland: A review of natural history and conservation requirements. In Olsen (ed.) (1993), pp. 196–208.

Debus, S.J.S. 1993. The status of the Red Goshawk *Erythrotriorchis radiatus* in New South Wales. In Olsen (ed.) (1993), pp. 182–191.

Debus, S.J.S. 1994. What is the Christmas Island Goshawk? *Australian Bird Watcher* 15: 377–379.

Debus, S.J.S. 1995a. Aerial display by Spotted Harrier *Circus assimilis*. *Australian Bird Watcher* 16: 167–168.

Debus, S.J.S. 1995b. Red Goshawk. *Nature Australia* 25(3):30–37.

Debus, S.J.S. 1996a. Further observations on the Square-tailed Kite. *Australian Birds* 29: 44–53; correction p. 62.

Debus, S.J.S. 1996b. Problems in identifying juvenile Square-tailed Kites. *Australian Bird Watcher* 16: 260–264.

Debus, S.J.S., McAllan, I.A.W. & Mead, D.A. 1993. Museum specimens of the Red Goshawk *Erythrotriorchis radiatus*. I. Annotated list of specimens; II. Morphology, biology and conservation status in eastern Australia. *Sunbird* 23: 5–28; 75–89.

Dekker, D. 1995. Prey capture by Peregrine Falcons wintering on southern Vancouver Island, British Columbia. *Journal of Raptor Research* 29: 26–29.

del Hoyo, J., Elliott, A. & Sargatal, J. 1994. *Handbook of the Birds of the World*, volume 2, *New World Vultures to Guineafowl*. Lynx, Barcelona.

Dennis, T.E. & Lashmar, A.F.C. 1996. Distribution and abundance of White-bellied Sea-Eagles in South Australia. *Corella* 20: 93–102.

Emison, W.B., Bren, W.M. & White, C.M. 1993. Influence of weather on the breeding of Peregrine Falcons *Falco peregrinus* near Melbourne. In Olsen (ed.) (1993), pp. 26–32.

Emison, W.B. & Hurley, V.G. 1995. Occupancy of Peregrine Falcon eyries near Melbourne during 1976–84 and 1992. *Victorian Naturalist* 112: 100–101.

Engel, D. & Rose, A.B. (forthcoming). Diet of the Black-shouldered Kite *Elanus axillaris* in New South Wales. *Australian Bird Watcher* 7(4).

Falkenberg, I.D., Dennis, T.E. & Williams, B.D. 1994. Organochlorine pesticide contamination in three species of diurnal raptor and their prey in South Australia. *Wildlife Research* 21: 163–173.

Garnett, S. (ed.) 1993. *Threatened and Extinct Birds of Australia*. Royal Australasian Ornithologists Union, Melbourne.

Gosper, D.G. 1994. Breeding of the Swamp Harrier on the NSW north coast. *Australian Birds* 27: 151.

Green, D.J. & Krebs, E.A. 1995. Courtship feeding in Ospreys *Pandion haliaetus*: A criterion for mate assessment? *Ibis* 137: 35–43.

Griffiths, C. 1994. Syringeal morphology and the phylogeny of the Falconidae. *Condor* 96: 127–140.

Harrison, F. & Lewis, M. 1997. Swamp Harriers breeding in north Queensland. *Australian Bird Watcher* 17: 102–103.

Helbig, A.J., et al. 1994. Phylogenetic relationships among falcon species (genus *Falco*) according to DNA sequence variation of the cytochrome b gene. In Meyberg & Chancellor (eds) (1994), pp. 593–600.

Holdaway, R.N. 1994. An exploratory phylogenetic analysis of the genera of the Accipitridae, with notes on the biogeography of the family. In Meyburg & Chancellor (eds) (1994), pp. 601–650.

Holdsworth, M. & Marmion, P. 1993. Raptors and education in Tasmania. In Olsen (ed.) (1993), pp. 220–226.

Hollands, D. 1984. *Eagles, Hawks and Falcons of Australia*. Nelson, Melbourne.

Hornsby, P. 1993. A comparison of the visual fields of various Australian raptors. In Olsen (ed.) (1993), pp. 317–328.

Hull, C. 1993. Prey dismantling techniques of the Peregrine Falcon *Falco peregrinus* and Brown Falcon *Falco berigora*: Their relevance to optimal foraging theory. In Olsen (ed.) (1993), pp. 330–336.

Hutton, K. 1994. Kleptoparasitism of Australian Ravens by Black Falcons. *Australian Birds* 28: 29–31.

Jenkins, A.R. 1995. Morphometrics and flight performance of southern African Peregrine and Lanner Falcons. *Journal of Avian Biology* 26: 49–58.

Kay, B.J., Twigg, L.E., Korn, T.J. & Nichol, H. 1994. The use of artificial perches to increase raptor predation on House Mice (*Mus domesticus*). *Wildlife Research* 21: 95–106.

Kenward, R.E. & Walls, S.S. 1994. The systematic study of radio-tagged raptors: I. Survival, home-range and habitat-use. In Meyburg & Chancellor (eds) (1994), pp. 303–316.

Lacombe, D., Bird, D.M., Hunt, K.A. & Mineau, P. 1994. The impact of fenthion on birds of prey. In Meyburg & Chancellor (eds) (1994), pp. 757–760.

Lavery, H.J. & Johnson, P.M. 1993. The Black Kite *Milvus migrans* in the Townsville district of Queensland: A comparison of irruption and non-irruption years. In Olsen (ed.) (1993), pp. 209–219.

Martin, G.R., Kirkpatrick, W.E., King, D.R., Robertson, I.D., Hood, P.J. & Sutherland, J.R. 1994. Assessment of the potential toxicity of the anticoagulant, pindone (2-pivalyl-1,3-indandione) to some Australian birds. *Wildlife Research* 21: 85–94.

Mathieson, M.T., Debus, S.J.S., Rose, A.B., McConnell, P.J. & Watson, K.M. 1997. Breeding diet of the Letter-winged Kite *Elanus scriptus* and Black-shouldered Kite *Elanus axillaris* during a House Mouse plague. *Sunbird* 27:65–71.

Meyburg, B.-U. & Chancellor, R.D. (eds) 1994. *Raptor Conservation Today*. WWGBP/Pica Press, London.

Meyburg, B.-U. & Chancellor, R.D. (eds). 1996. *Eagle Studies*. World Working Group on Birds of Prey and Owls, Berlin.

Meyburg, B.-U., et al. 1996. Satellite tracking of eagles: Method, technical progress and first personal experiences. In Meyburg & Chancellor (eds) (1996), pp. 529–549.

Mooney, N. (ed.). 1994. Rehabilitation special. *Australasian Raptor Association News* 15(2/3): 23–68.

Mooney, N.J. & Brothers, N. 1993. Dispersion, nest and pair fidelity of Peregrine Falcons in Tasmania. In Olsen (ed.) (1993), pp. 33–42.

Mooney, N.J. & Taylor, R.J. 1996. Value of nest site protection in ameliorating the effects of forestry operations on Wedge-tailed Eagles in Tasmania. In Bird, D.M., Varland, D.E. & Negro, J.J. (eds) (1996), pp. 275–282.

Newgrain, K., Green, B., Olsen, P., Mooney, N. & Brothers, N. 1993. Validation of 22-sodium turnover for the estimation of food consumption and energy requirements of some captive Australian raptors. In Olsen (ed.) (1993), pp. 285–294.

Newgrain, K., Olsen, P., Green, B., Mooney, N., Brothers, N. & Bartos, R. 1993. Food consumption rates of free-living raptor nestlings. In Olsen (ed.) (1993), pp. 274–284.

Olsen, J. 1994a. *Caring for Birds of Prey*. Wild Ones, Springville, CA.

Olsen, J. 1994b. *Some Time with Eagles and Falcons*. University of Canberra, Canberra.

Olsen, J. & Georges, A. 1993. Do Peregrine Falcon fledglings reach independence during peak abundance of their main prey? *Journal of Raptor Research* 15: 149–153.

Olsen, J., Olsen, P., Billett, T. & Jolly, J. 1993. Observations on the diet of the Peregrine Falcon in South Australia. In Olsen (ed.) (1993), pp. 81–83.

Olsen, J. & Stevenson, E. 1996. Female Peregrine Falcon *Falco peregrinus* replaces an incubating female and raises her young. *Australian Bird Watcher* 16: 205–210.

Olsen, P. 1995. *Australian Birds of Prey*. New South Wales University Press, Sydney.

Olsen, P. (ed.). 1993. *Australian Raptor Studies*. Australasian Raptor Association, RAOU, Melbourne.

Olsen, P. & Allen, T. 1997. The trials of quarry-nesting Peregrine Falcons. *Australian Bird Watcher* 17:87–90.

Olsen, P., Crome, F. & Olsen, J. 1993. *Birds of Prey and Ground Birds of Australia*. Angus & Robertson, Sydney.

Olsen, P., Morris, A. & Bennett, C. 1993. Observations on the breeding diet of the Peregrine Falcon in New South Wales. In Olsen (ed.) (1993), pp. 78–80.

Olsen, P., Olsen, J. & Mason, I. 1993. Breeding and non-breeding season diet of the Peregrine Falcon *Falco peregrinus* near Canberra, prey selection, and the relationship between diet and reproductive success. In Olsen (ed.) (1993), pp. 55–77.

Olsen, P.D. & Cockburn, A. 1993. Do large females lay small eggs? Sexual dimorphism and the allometry of egg and clutch volume. *Oikos* 66: 447–453.

Olsen, P.D., Fuller, P. & Marples, T.G. 1993. Pesticide-related eggshell thinning in Australian raptors. *Emu* 93: 1–11.

Olsen, P.D. & Marples, T.G. 1993. Geographic variation in egg size, clutch size and date of laying of Australian raptors (Falconiformes and Strigiformes). *Emu* 93: 167–179.

Oro, D. & Tella, J.L. 1995. A comparison of two methods for studying the diet of the Peregrine Falcon. *Journal of Raptor Research* 29: 207–210.

Peeters, H.J. 1994. Suspected poisoning of Golden Eagles *Aquila chrysaetos* by chloropha-cinone. In Meyberg & Chancellor (eds) (1994), pp. 775–776.

Pettigrew, J.D. 1993. Some observations on avian visual optics, with special reference to the bifoveate diurnal birds of prey. In Olsen (ed.) (1993), pp. 299–316.

Ragless, G.B. 1995. Unusual nest sites of the Wedge-tailed Eagle *Aquila audax*. *South Australian Ornithologist* 32: 61.

Reymond, E. 1993. Spatial visual acuity of three Australian raptors [summary only]. In Olsen (ed.) (1993), p. 329.

Schodde, R. 1993. Origins and evolutionary radiations in Australia's birds of prey [summary only]. In Olsen (ed.) (1993), p. 12.

Seibold, I., Helbig, A.J. & Wink, M. 1993. Molecular systematics of falcons (family Falconidae). *Naturwissenschaften* 80: 87–90.

Seibold, I., et al. 1996. Genetic differentiation and molecular phylogeny of European *Aquila* eagles according to cytochrome b nucleotide sequences. In Meyburg & Chancellor (eds) (1996), pp. 1–16.

Simmons, R. & Mendelsohn, J. 1993. A critical review of cartwheeling flights of raptors. *Ostrich* 64: 13–24.

Stokes, T. 1996. Helicopter effects upon nesting White-bellied Sea-Eagles and upon smaller birds at an isolated protected location (Eshelby Island, Great Barrier Reef, Australia). *Corella* 20: 25–28.

Thiollay, J.-M. 1993. Habitat segregation and the insular syndrome in two congeneric raptors in New Caledonia, the White-bellied Goshawk *Accipiter haplochrous* and the Brown Goshawk *A. fasciatus*. *Ibis* 135: 237–246.

Twigg, L.E. & Kay, B.J. 1994. Changes in the relative abundance of raptors and House Mice in western New South Wales. *Corella* 18: 83–86.

Viñuela, J. 1996. Establishment of mass hierarchies in broods of the Black Kite. *Condor* 98: 93–99.

Viñuela, J., Villafuerte, R. & de le Court, C. 1994. Nesting dispersion of a Black Kite population in relation to location of rabbit warrens. *Canadian Journal of Zoology* 72: 1680–1683.

Walls, S.S. & Kenward, R.E. 1994. The systematic study of radio-tagged raptors: II. Sociality and dispersal. In Meyburg & Chancellor (eds) (1994), pp. 317–324.

White, C.M., Parrish, J.R., Brimm, D.J. & Longmire, J.L. 1993. Aspects of variation between Peregrine Falcon populations: A review with emphasis on the Southern Hemisphere. In Olsen (ed.) (1993), pp. 13–14.

Wilson, S.C. & Whelan, R.J. 1993. Prey camouflage in the Sugar Glider and predatory response by the Variable Goshawk. In Olsen (ed.) (1993), pp. 337–343.

Young, J. & De Lai, L. 1997. Population declines of predatory birds coincident with the introduction of Klerat rodenticide in North Queensland. *Australian Bird Watcher* 7: 160–167.